INTENSIFICATION OF SORPTION PROCESSES

INTENSIFICATION OF SORPTION PROCESSES

ACTIVE AND PASSIVE MECHANISMS

MAHMOOD REZA RAHIMI

Professor, Process Intensification Laboratory, Chemical Engineering Department, Engineering School, Yasouj University, Yasouj, Iran

SOLEIMAN MOSLEH

Assistant Professor, Polymer Engineering Department, Faculty of Gas and Petroleum, Yasouj University, Gachsaran, Iran

ELSEVIER

Elsevier
Radarweg 29, PO Box 211, 1000 AE Amsterdam, Netherlands
The Boulevard, Langford Lane, Kidlington, Oxford OX5 1GB, United Kingdom
50 Hampshire Street, 5th Floor, Cambridge, MA 02139, United States

Notices

Knowledge and best practice in this field are constantly changing. As new research and experience broaden our understanding, changes in research methods, professional practices, or medical treatment may become necessary.

Practitioners and researchers must always rely on their own experience and knowledge in evaluating and using any information, methods, compounds, or experiments described herein. In using such information or methods they should be mindful of their own safety and the safety of others, including parties for whom they have a professional responsibility.

To the fullest extent of the law, neither the Publisher nor the authors, contributors, or editors, assume any liability for any injury and/or damage to persons or property as a matter of products liability, negligence or otherwise, or from any use or operation of any methods, products, instructions, or ideas contained in the material herein.

Library of Congress Cataloging-in-Publication Data
A catalog record for this book is available from the Library of Congress

British Library Cataloguing-in-Publication Data
A catalogue record for this book is available from the British Library

ISBN: 978-0-12-821411-4

Publisher: Susan Dennis
Acquisitions Editor: Anita Koch
Editorial Project Manager: Ivy Dawn Torre
Production Project Manager: Sruthi Satheesh
Cover Designer: Victoria Pearson

Typeset by TNQ Technologies

Contents

Preface

Process intensification (PI) involves innovative processing techniques, materials, and equipment, often based on alternative forms of energy input aimed at enhancing process performance and at reducing capital and/or operating costs. The main objective of PI-based technologies is the reduction in capital cost of an existing production system deriving from physical miniaturization of process equipment and/or the application of innovative separation techniques. Designing novel equipment for reduction of capital and operational costs, and improving both mass transfer and heat transfer, are some of the main challenges for engineers and scientists, especially in the sorption processes.

This book introduces a number of selected, advanced topics in sorption processes/PI, covering both theoretical and applicable aspects. This book includes a full-fledged look at the sorption processes in terms of the basic principles and design, details of materials or equipment, industrial upscaling, economic evaluation/justification, future opportunities, and challenges for each sorption process. The first part of the book is devoted to the study of sorption processes based on active mechanisms, including ultrasonic, microwave, high-gravity, electrical, and magnetic fields, while the second part covers passive mechanisms like nanostructures and nanofluids, membrane, supercritical fluids, and sorption processes based on geometry design and equipment structure. The focus of the book is on key aspects of novel PI technologies (processes and equipment), i.e., absorption and adsorption, working principles, and design and applications. This book introduces the active and passive mechanisms in PI-based processes and equipment. Design, fabrication, and optimization of different innovative sorption processes that operate based on both mentioned mechanisms are well discussed.

This book was written at a time when the world was combating the spread of the COVID-19 epidemic. Although this epidemic had a very bad impact on all of us emotionally, it has brought with it some tips for the future. First, it showed how the decision to control an epidemic in any part of the world affects the whole world in such cases. And second, it showed how paying attention to reliable technologies can help to respond appropriately in the shortest possible time and dramatically reduce the idea cycle to the end product. As we have seen in the production of the COVID-19 vaccine, this pathway will be useful in the future. PI is one of the basic frameworks needed for the future, and more attention should definitely be paid to its role in all dimensions. In this regard, we have dealt with this issue briefly in the end part of the introduction.

The main audiences for this book include graduate and postgraduate students, researchers in academia and industry, and chemical and environmental engineers working in the field of separation processes. The book is also intended for the designing of process and equipment related to different industries, including gas separation processes, wastewater treatment, and the removal of contaminants. We hope that this book can meet the expectations of engineers and researchers in the field of separation, especially the processes involved in sorption, and be an important step in the application and development of innovative processes, materials, and equipment.

Mahmood Reza Rahimi
Soleiman Mosleh

Introduction

1 Significance of process intensification

Process intensification (PI) provides innovative equipment, materials, and processes that lead to remarkable enhancements in manufacturing and processing, significantly decreasing equipment-size/production-capacity ratio, energy consumption, or waste production, and eventually bring cheaper and sustainable technologies (Bielenberg & Palou-Rivera, 2019). In fact, PI removes the existing restrictions by enhancing mass, heat, and momentum transfer rates (Demirel, Li, & Hasan, 2019). A dramatic reduction in processing cost could be occurred by utilizing the synergy among the multifunctional phenomena, while ensuring feasibility, operability, controllability, and safety (Baldea & Edgar, 2018). PI-based processes are inherently safer, mainly owing to the smaller size and volume of equipment/processes, and subsequently less quantity of toxic substances (Ebrahimi, Virkki-Hatakka, & Turunen, 2012). PI includes a wide range of tools, which are classified into three parts including intensified equipment, intensified methods, and advanced materials (Fig. 1), as follows (Abdulrahman, Máša, & Teng, 2020):

- Intensified equipment: designing and fabrication of devices based on innovative configurations that can significantly improve the rates of mass, heat, and momentum transfer.
- Intensified methods: utilizing novel techniques to intensify processes including incorporating separation processes or chemical reactions into unit operations. In this method, a synergistic effect will be achieved through the combination of some individual techniques, which lead to the enhancement of unit performance.
- Advanced materials: PI provides sustainable technologies for the synthesis and production of novel materials with excellent properties. PI enables tuning the structure of material during the novel synthesis methods. This means that various characterizations can be engineered such as particle size distribution and morphology.

Another category of PI technologies is presented in Fig. 2. This category is provided by consideration of the unit/process intensification.

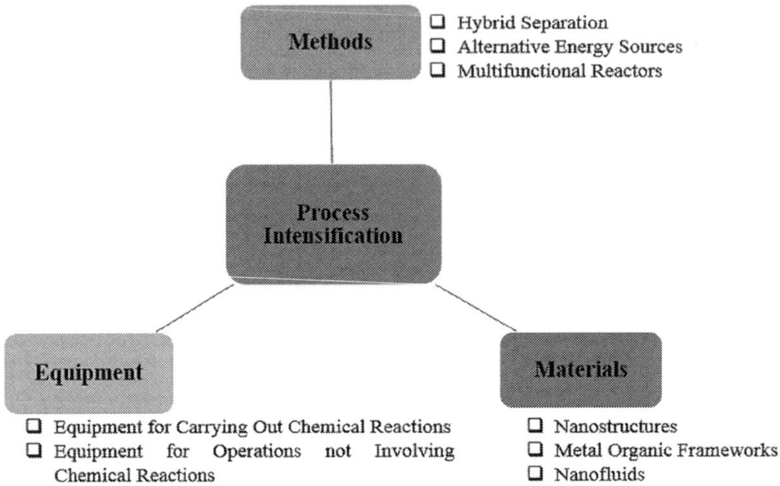

Fig. 1 Classification of PI tools.

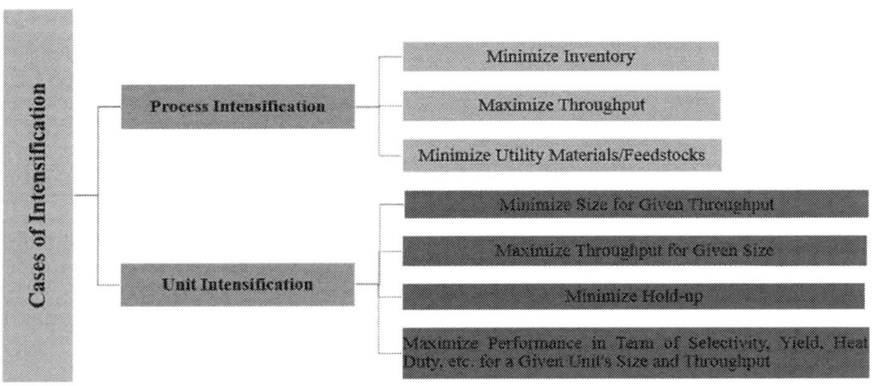

Fig. 2 Classification of PI technologies based on the process/unit intensification (Ponce-Ortega, Al-Thubaiti, & El-Halwagi, 2012).

Fig. 3 shows an example of PI technology on a commercial scale for the production of methyl acetate. As can be seen, seven tasks have been integrated into a single piece of equipment. This opportunity not only cause a decrease in costs but also leads to saving energy.

From the point of view of chemical engineering, PI tries to presents innovative, sustainable, and efficient techniques for the manufacturing and processing of chemical products. In this regard, different strategies can vary applied depending on the field of chemical engineering, such as process

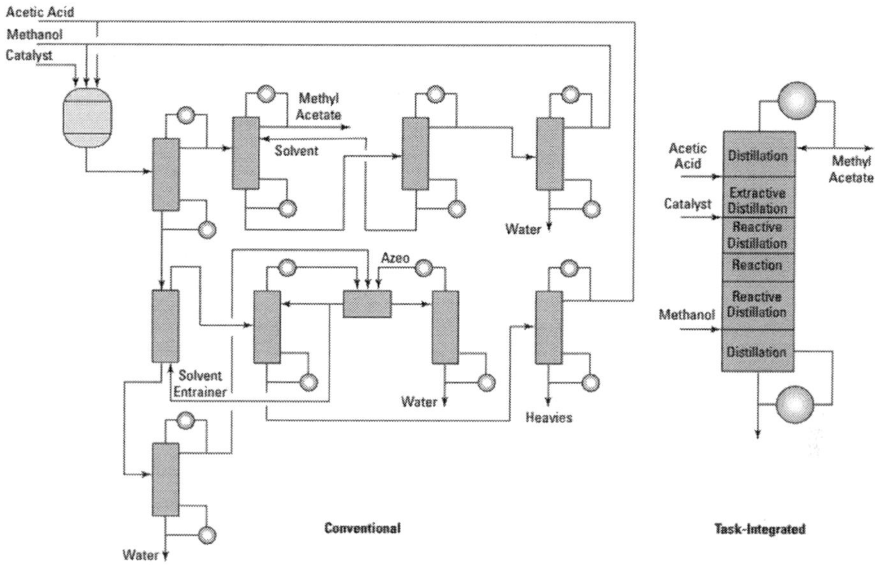

Fig. 3 Methyl acetate production: PI technology versus conventional method (Stankiewicz & Moulijn, 2000).

system engineering (PSE). Four main domains have been identified including structure (spatial domain), energy (thermodynamic domain), synergy (functional domain), and time (temporal domain). Table 1 provides details of these main domains.

Table 1 Main domains of PI technology (Keil, 2018).

Structure: spatial domain	**Energy: thermodynamic domain**	**Synergy: functional domain**	**Time: temporal domain**
Structure in molecular events	Bringing energy to molecules (what form and how)	Synergy on molecular scale	Timing of events
Structure in catalysts	Bringing energy to catalysts	Synergy in transport processes	Applying dynamics
Structure in phase contacting	Energy transfer in hydrodynamics, mixing, and transport processes	Synergy in processing units—multifunctional reactors and separators	Special process control
Structure in transport phenomena	Energy management in reactors and separation systems		

PI-based technologies impose the following approaches at different scales, from the molecular processes, passing through the microscale (microfluidic devices), to macroscale (reactors and columns), and up to the megascale (units, plants, and sites) (Portha, Falk, & Commenge, 2014; Rivas, Castro-Hernández, Perales, & van der Meer, 2018):

- Maximizing the effectiveness of intra- and intermolecular events.
- Giving each molecule the same processing experience.
- Optimizing the driving forces and maximizing the specific areas to which these forces apply.
- Maximizing synergistic effects from partial processes.

2 PI-involved mechanisms: active and passive

The enhancement mechanisms of PI technology are classified in mainly two basic ways: active and passive. Passive refers to such techniques that do not use any external actuator to drive the fluids, guide the particles in the fluid, separate them, and so forth. In contrast, active techniques utilize external power (pressure filed, acoustic filed, microwave field, magnetic field, electric field, thermal field, and so forth) for processes (Bayareh, Ashani, & Usefian, 2020). Although both active and passive techniques are applied to intensify heat and mass transfer, each of them has some advantages and disadvantages. The use of external energy sources provides a great environment for intensification of the mass and heat transfer, but it leads to an increase in costs. The passive techniques do not require external energy, which makes them affordable, but passive-based devices have a complex structure and this may cause several limitations during the design and fabrication steps. In general, it can be concluded that in passive techniques, the "structure" approach plays the primary role, while in the active techniques "energy" (acoustic, electrical, magnetic, and so forth) or "time" (pulsing and oscillations) become dominant (Stankiewicz, Van Gerven, & Stefanidis, 2019). The treated surfaces, rough surfaces, extended surfaces, displaced enhancement devices, swirl flow devices, surface tension devices, porous structures, additives, coiled tubes, and surface catalysis are some examples of passive-based techniques, whereas the mechanical aids, surface vibration, fluid vibration, electrostatic fields, other electrical methods, suction or injection, jet impingement, rotation, induced flow instabilities, and grooves and rivulets fall into the active category (Reay, Ramshaw, & Harvey, 2013).

3 Intensification of the sorption processes

Sorption processes as the most common separation in chemical engineering that have a key role in the production of wide-ranging products. In this regard, focusing on novel technologies for the production of high-quality chemicals with low energy consumption and minimal impact on the environment has attracted a lot of attention. Great challenges still remain in many sorption processes owing to their high share in total energy demand. It is estimated that energy used for separation processes (including sorption) is about half of US industrial energy use, 10%–15% of the nation's total energy consumption, and about 60% of the total energy required by the chemical plants (Tian, Demirel, Hasan, & Pistikopoulos, 2018). For example, distillation has a high share of energy consumption of the total energy used in separation (about 50%) because of its relatively high energy intensity. PI can intensify the sorption processes to achieve high performance with minimal energy consumption through designing of innovative equipment/processes such as hybrid nonreactive separation (e.g., dividing wall column, membrane distillation, and heat-integrated distillation), combined reaction/separation (e.g., reactive distillation and reactive extraction), and using external fields (e.g., high-gravity, electrical, magnetic, ultrasound, and microwave). Table 2 presents the application of main PI-based sorption technologies including hybrid nonreactive separation (HNS), membrane-assisted separation (MNS), and periodic operations (PO). As can be seen, dividing wall column (DWC) and MNS are still the most preferred technologies for recovery processes in petrochemical industries. The main applications of the aforementioned technologies are water and wastewater treatments, gas processing, recovery processes in the petrochemical plants, and pharmaceutical industry.

PI technologies have excellent potential for carbon capture. Energy savings, high carbon capture efficiency, maturity, and probability of overcoming limitations are great benefits of PI. Several PI-based technologies which can be used for CO_2 capture with high performance, are listed in Table 3.

Table 2 PI-based sorption technologies in the chemical industry (Abdulrahman et al., 2020).

		Separation technologies			
PI Tech.	**Main technologies/ configuration**	**Commercial-scale application**	**The main field of application**	**Implement action Degree [x times] or [%]**	**Reference**
HNS	• Side strippers • Side rectifiers • Fully coupled which was modified later • Dividing wall column	• Dividing wall column (DWC)	• Petrochemical	>125	Asprion & Kaibel (2010)
MNS	• Pervaporation (PV) • Vapor Permeation(VP) • Micro-Filtration • Nano-Filtration • Reverse Osmosis • Membrane absorption • Membrane distillation	• Reverse osmosis (RO)	• Water treatment • Pharmaceutical • CO_2 capturing • Petrochemical (in recovery processes)	60% of worldwide plants	Buonomenna (2013), Lee, Arnot, and Mattia (2011), Lutze and Gorak, 2013
PO	• Pressure swing adsorption (PSA) • Temperature swing adsorption (TSA) • Simulated moving bed (SMB) • Batch distillation	• Pressure swing adsorption (PSA) • Simulated moving bed (SMB) • Continuous mode adsorption	• Petrochemical • Dehydration • CO_2 separation • Hydrogen purification • Methane purification Air separation • Petrochemical (mainly refinery) • Sugar plants		Broughton and Gerhold (1961), Storti, Baciocchi, Mazzotti, and Morbidelli (1995), Voss (2005)

Table 3 PI-based technologies for CO_2 reduction (Pereira & Patel, 2019).

Technology	Potential for energy savings	Potential to improve cost competitiveness	Potential to reduce CO_2	Maturity of technology	Likeliness of overcoming barriers
Heat-integrated distillation/DWC	High	High	High	High	High
Reactive distillation	High	High	High	High	High
Membrane-assisted reactive distillation	High	High	High	High	Medium
Static mixer reactors for continuous reactions	High	Medium	High	High	High
Pulsed compression reactor	High	Medium	High	Low	Low
Reactive absorption	High	Low	High	High	High
Supercritical separations	Medium	High	High	Medium	High
Reverse flow reactor operations	Medium	Medium	High	Medium	High

4 PI technology and its role in the control of the COVID-19 pandemic

One of the main benefits of PI-based technology is the ability to work in the field of the biochemical and pharmaceutical industries. In pandemic conditions such as the COVID-19 outbreak, the significance of the PI-based technology becomes more apparent. The pandemic could aggravate food insecurity and restricts public access to (bio)chemicals and pharmaceutical drugs, and vaccines. This situation could push a large number of households into the poverty trap, resulting in malnutrition and death (Sampath, Jagadeesh, & Bahinipati, 2020). Overcoming these challenges will be necessary in such a crisis situation. On the other hand, due to the disease pandemic such as COVID-19, global competition in the chemical, biochemical, and pharmaceutical industry is continuously increasing. Under this condition, the development of sustainable technologies for the reduction of process development time and investment costs is a crucial step (Buchholz, 2010). A great challenge, especially in the epidemic time of diseases such as COVID-19, is the inequities inherent in vaccine production. Ensuring global access to an effective vaccine is the main objective of some organizations such as the World Health Organization (WHO). The production of enough vaccines to meet demand will require manufacturing innovation. The vaccine production is not only expensive but also is has a typically slow manufacturing rate. Furthermore, there are some barriers during the design, build, validation, and commence commercial manufacturing. PI has an inherent potential to provides innovative techniques to reduce the time and space required to make vaccines, while also reducing processing complexity. Besides, PI can decrease operator-dependent risks, which makes it proper for countries lacking the workforce skills required to run traditional plants (Anderson, 2020). For instance, a 50-liter bioreactor has been made using a three-dimensional matrix of polyethylene fibers that can operate as much as a traditional 1000-liter bioreactor. The cost-effective reactors can be designed for the production of various vaccines. A platform that combines chaining (continuous or semi-continuous processing) and automation, enabling manufacture with an extremely reduced footprint. For example, a facility with a footprint of 6 m^2 can replace conventional devices that would take up 120 m^2 (Fig. 4).

Fig. 4 A small-footprint bioreactor for intensifying vaccine production (Univercells/ Didier Ropers) (Anderson, 2020).

References

Abdulrahman, I., Máša, V., & Teng, S. Y. (2020). Process intensification in the oil and gas industry: A technological framework. *Chemical Engineering and Processing: Process Intensification*, 108208.

Anderson, T. (2020). Intensifying vaccine production, World Health Organization. *Bulletin of the World Health Organization, 98*, 302–303.

Asprion, N., & Kaibel, G. (2010). Dividing wall columns: fundamentals and recent advances. *Chemical Engineering and Processing: Process Intensification, 49*, 139–146.

Baldea, M., & Edgar, T. F. (2018). Dynamic process intensification. *Current Opinion in Chemical Engineering, 22*, 48–53.

Bayareh, M., Ashani, M. N., & Usefian, A. (2020). Active and passive micromixers: a comprehensive review. *Chemical Engineering and Processing: Process Intensification, 147*, 107771.

Bielenberg, J., & Palou-Rivera, I. (2019). The RAPID Manufacturing Institute—Reenergizing US efforts in process intensification and modular chemical processing. *Chemical Engineering and Processing: Process Intensification, 138*, 49–54.

Broughton, D. B., & Gerhold, C. G. (1961). *Continuous sorption process employing fixed bed of sorbent and moving inlets and outlets*. Google Patents.

Buchholz, S. (2010). Future manufacturing approaches in the chemical and pharmaceutical industry. *Chemical Engineering and Processing: Process Intensification, 49*, 993–995.

Buonomenna, M. G. (2013). Membrane processes for a sustainable industrial growth. *RSC Advances, 3*, 5694–5740.

Demirel, S. E., Li, J., & Hasan, M. F. (2019). Systematic process intensification. *Current Opinion in Chemical Engineering, 25*, 108–113.

Ebrahimi, F., Virkki-Hatakka, T., & Turunen, I. (2012). Safety analysis of intensified processes. *Chemical Engineering and Processing: Process Intensification, 52*, 28–33.

Keil, F. J. (2018). Process intensification. *Reviews in Chemical Engineering, 34*, 135–200.

Lee, K. P., Arnot, T. C., & Mattia, D. (2011). A review of reverse osmosis membrane materials for desalination—Development to date and future potential. *Journal of Membrane Science, 370*, 1–22.

Lutze, P., & Gorak, A. (2013). Reactive and membrane-assisted distillation: Recent developments and perspective. *Chemical Engineering Research and Design, 91*, 1978–1997.

Pereira, C. S., & Patel, B. A. (2019). The role of process intensification in addressing the dual energy challenge. *Chemical Engineering and Processing: Process Intensification, 142*, 107545.

Ponce-Ortega, J. M., Al-Thubaiti, M. M., & El-Halwagi, M. M. (2012). Process intensification: New understanding and systematic approach. *Chemical Engineering and Processing: Process Intensification, 53*, 63–75.

Portha, J.-F., Falk, L., & Commenge, J.-M. (2014). Local and global process intensification. *Chemical Engineering and Processing: Process Intensification, 84*, 1–13.

Reay, D., Ramshaw, C., & Harvey, A. (2013). *Process Intensification: Engineering for efficiency, sustainability and flexibility*. Butterworth-Heinemann.

Rivas, D. F., Castro-Hernández, E., Perales, A. L. V., & van der Meer, W. (2018). Evaluation method for process intensification alternatives. *Chemical Engineering and Processing: Process Intensification, 123*, 221–232.

Sampath, P. V., Jagadeesh, G. S., & Bahinipati, C. S. (2020). Sustainable intensification of agriculture in the context of the COVID-19 pandemic: Prospects for the future. *Water, 12*, 2738.

Stankiewicz, A. I., & Moulijn, J. A. (2000). Process intensification: Transforming chemical engineering. *Chemical Engineering Progress, 96*, 22–34.

Stankiewicz, A., Van Gerven, T., & Stefanidis, G. (2019). *The fundamentals of process intensification*. John Wiley & Sons.

Storti, G., Baciocchi, R., Mazzotti, M., & Morbidelli, M. (1995). Design of optimal operating conditions of simulated moving bed adsorptive separation units. *Industrial and Engineering Chemistry Research, 34*, 288–301.

Tian, Y., Demirel, S. E., Hasan, M. F., & Pistikopoulos, E. N. (2018). An overview of process systems engineering approaches for process intensification: State of the art. *Chemical Engineering and Processing: Process Intensification, 133*, 160–210.

Voss, C. (2005). Applications of pressure swing adsorption technology. *Adsorption, 11*, 527–529.

SECTION ONE

Active mechanisms

CHAPTER ONE

Ultrasonic and microwave-assisted sorption processes

1.1 Introduction

Conventional intensification techniques for mass transporting purposes are not always applicable owning to several unwanted side effects (Li, Wu, Hao, Li, & Gao, 2019). Ultrasonic and microwaves are green irradiation methods applied frequently in different fields of science include separation and mass transferring domains (Gogate, 2017). Ultrasound waves produced acoustic cavitation in the proposed sample leading to physical forces which are regarded as the main reason responsible for alterations in exposed materials. Microwaves are a region between infrared and radio waves in the electromagnetic spectrum (Martin & Navarrete, 2018). These two methods are appeared resolving different challenges lead to solvent free and increased process yields. Ultrasound is a well-known and broadly used method in processing and synthesis domains providing unique and beneficial operating conditions (Holkar, Jadhav, Pinjari, & Pandit, 2019). Power ultrasound is regarded as a promising approach to speed up mass transferring mechanism in textile materials.

Sonochemistry is based on two major principles as follows:

(a) Mechanical practices for mixing and disintegration purpose.
(b) High-energy processing for radical based reactions.

The ultrasonic was occurred mostly in liquids, where sound waves are able to disrupt the continuum (Chipurici et al., 2019). Cavitation is produced in shock tubes caused by oscillating bubbles, pumps, and other hydrodynamic equipment. Due to the applied high frequencies in sonochemistry between 16 kHz and up to megahertz domains, drastic conditions created by the result. Radiation forces create intense micro- and macromixing with high shear forces, which are used in emulsification, homogenization, and fragmentation processes (Abd Malek et al., 2020).

Intensification of Sorption Processes
ISBN: 978-0-12-821411-4
https://doi.org/10.1016/B978-0-12-821411-4.00004-3

Produced microjets and shockwaves, led to cleaning, dispersion, activation, and fragmentation of solid materials. Besides the main industrial usage, other applications, such as solids processing, atomization, crystallization, environmental protection, and separation can be mentioned to be used.

The term ultrasound refers the waves in a frequency of 16 kHz up to megahertz domain. Vibrational frequencies are transferred to fluid and via oscillating devices led to produce pressure waves (Yusof, Shariff, Tay, Lau, & Mustafa, 2020). Liquids are disturbed using sound assisted pressure led to create so-called cavities as a result. Most sonochemical phenomena are obtained by oscillating bubbles are considered as secondary effects.

Ultrasonication is a promising and efficient mechanical pretreatment approach in different fields of science. The ultrasound methods can be described as a cyclic sound pressure (compression and expansion) in frequencies higher than 20 kHz (Guo, Zhang, Li, Wang, & Kougoulos, 2005; Park et al., 2007). The applications and range of ultrasound waves can be seen in Fig. 1.1.

Depending on the applied frequency range, it is divided into three different regions as following: power ultrasound (ranged between 20 and 100 kHz), high frequency ultrasound (ranged between 100 kHz and 1 MHz), and diagnostic ultrasound (ranged between 1 and 500 MHz). With developments in technology and industries, ultrasound domain (>20 kHz) is used in several fields (Park & Yeo, 2008). The high frequency ultrasound between 1 and 10 MHz is employed in various fields, such as animal navigation, determination of cracks or flaws in solid samples and echolocation application under water, quantification of pelvic abnormalities, curing benign and cancer tissues tumors, and so forth.

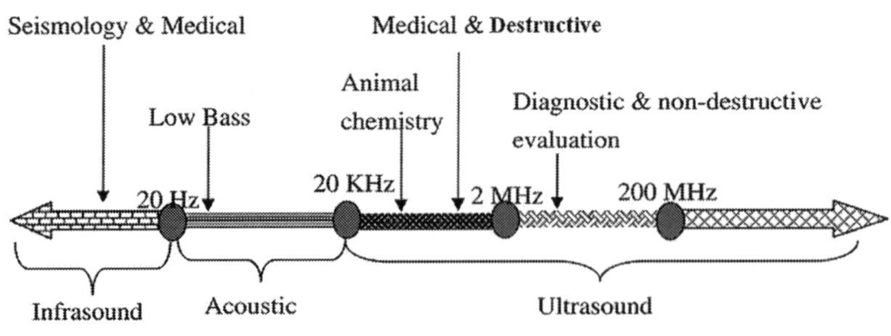

Figure 1.1 Diagram of ultrasound range and its different applications.

1.2 Mechanism of the ultrasonic activation

Cavitation happens followed by sonication, which simultaneously causes mechanical, chemical and physical phenomena (Vázquez-López et al., 2019). Each of these phenomena traces the following principle:

- Mechanical

The negative and positive pressure zones are caused by ultrasonic waves in the medium. A bubble is made for the lower pressure with time-based critic or unstable size resulting in its implosion and increasing the cutting forces (shear forces).

- Physical

The medium's temperature is increased by the continuous implosion of bubbles and the cavitation-caused shear forces.

- Chemical

Sonochemical reactions are made in the aqueous medium leading to the creation of H• and OH• radicals and oxidation of intracellular compounds and cell wall. H_2O_2 and H_2O are the end products of the oxidation reactions that are not considered toxic and the former one is still an oxidizing compound (Borea et al., 2018).

The ability obtained via ultrasound domain in triggering and/or speeding up many chemical reactions is a promising field of science owning to the opportunities offered to produce and synthesize new products and compound with better yields and in shorter periods of time. Sonochemical phenomena created by interaction occurs between a proper acoustic waves range and a highly reactive chemical system; by the way, the interaction performed via the intermediate phenomenon of the acoustic cavitation. It should be noted that three important factors should be considered in investigating a sonochemical reaction which can be mentioned as following: the chemical system, the acoustic field, and the bubbles' field. Fig. 1.2 indicates a schematic representation of ultrasonication process (Pilli et al., 2011).

Chemical systems consist of a solution containing reactants, the products and, also, unstable species (radicals). The liquid medium may be homogeneous which means only one phase to be presented. Also, the liquid can be characterized by some physical parameters, include viscosity, surface stress, temperature, density. When the reactants are not dissolved into a solution or when some solid particles are produced, a more complex physical conditions occurs. The acoustic field consists of acoustic waves moving

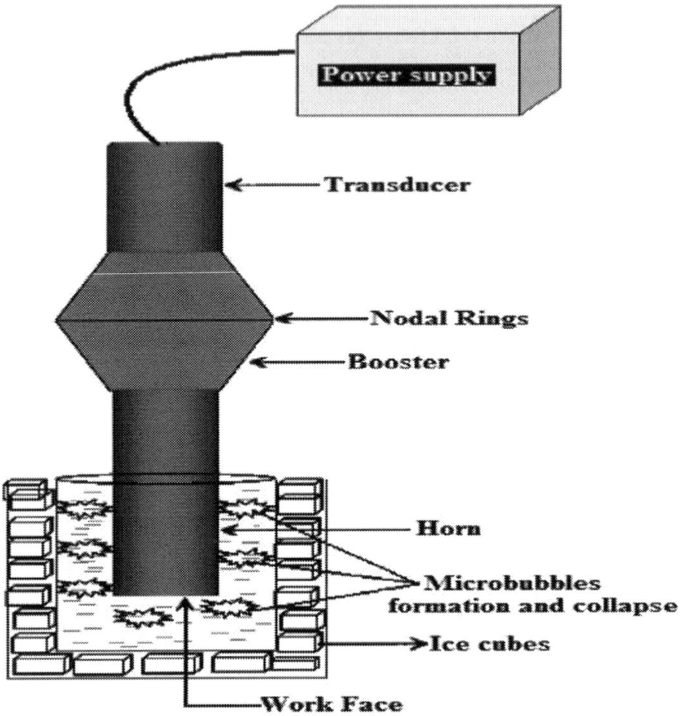

Figure 1.2 An example of ultrasonication process (Pilli et al., 2011).

through the liquid media. The bubbles' field include some components of empty and gas (or vapor) filled cavities, which are located in the liquid medium which was appeared after being created by several mechanisms (Pilli et al., 2011).

1.3 Mechanism of the microwave irradiation

Microwaves (MW) are a region of electromagnetic waves which take up in the range between infrared radiation (IR) and radio waves in the spectrum. Compared to traditional heating approaches such as hotplate heating technique, microwave irradiation (MWI) possesses several benefits include remarkably reduced reaction time, the capability of more rapid product creation, low reaction activation energy (Motevali, Minaei, Banakar, Ghobadian, & Khoshtaghaza, 2014). The aforementioned benefits have introduced MWI as a desired technique in several fields including biomass. Fig. 1.3 illustrated a dynamic microwave-assisted pretreatment system.

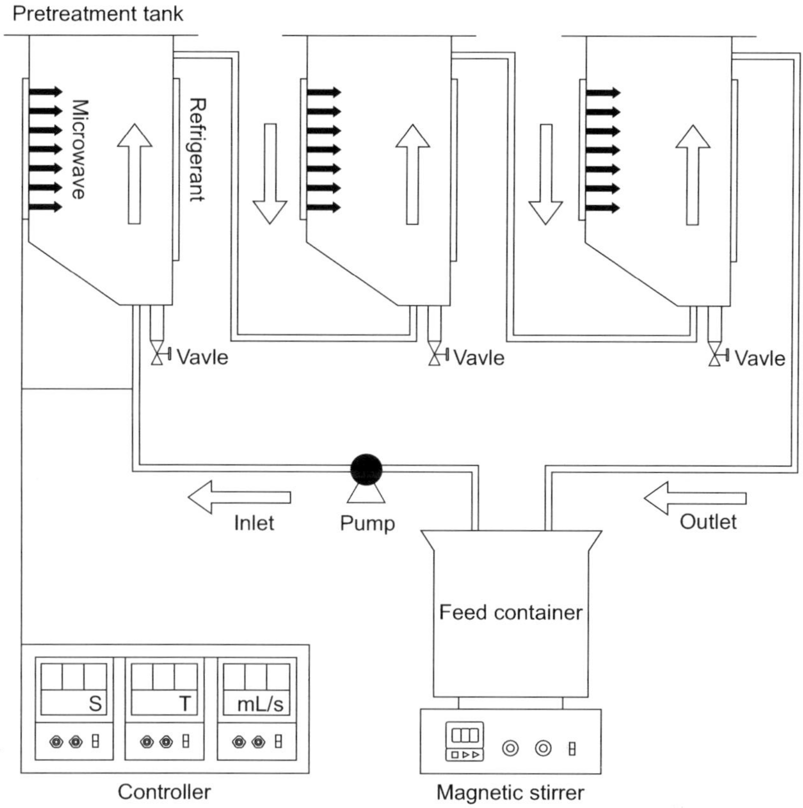

Figure 1.3 Dynamic microwave-assisted pretreatment system (Li, Qu, Yang, Chang, & Xu, 2016).

There is evidence say that microwaves can lead to different biological compacts depending some indicators such as the field strength, frequencies, modulation, and exposure duration (Rai, Singh, Mishra, Singh, & Samarketu, 2009). These compacts were mainly objected to heating property of microwaves (Pakhomov, Prol, Mathur, Akyel, & Campbell, 1997), but recent studies have represented that some nonthermal effects of microwaves. One of the recently applied methods in food sterilization domain is the employ of microwaves. Microbiological researches used microwave irradiation possesses some problems lead to cell death. One of the major reasons for the mentioned problems is the difficulty of temperature to be kept constant during the irradiation.

Another area that attract several attentions of microwaves is biomass. Biomass possesses the potential of becoming an excellent global energy source due to the ability of making a deal to response to future energy demands (Bauen et al., 2009). There are numerous bioenergy methods that can be employed to convert the raw feedstocks into accessible energy and chemical materials which can be divided into two major classes include thermochemical and biochemical conversion ways. Fig. 1.4 represents the major bioenergy conversion pathways that have already been used successfully (Kostas, Beneroso, & Robinson, 2017).

One of the main fields of microwave domains is in heating-assisted methods. The traditional approaches, however, have several drawbacks as follows: (1) heat losses and low-quality steams in deep formation aspects; (2) the difficult penetrating capability of injected steam into the reservoir; (3) early injected fluid breakthrough and declined sweep; and (4) commercial problems. In this regard, some techniques included microwaves can be a promising tool for in situ heating of the reservoirs. Fig. 1.5 illustrates a microwave-assisted heating system.

According to the past studies, it can be understood that the microwave possesses the ability to radiate in the borehole via the antenna. Furthermore, as increasing the temperature, the sorption ability of coal is reduced as a result (Bae & Bhatia, 2006). In addition, the microwave energy has the ability to duce fractures and rise cleat apertures under tension (Kumar et al., 2011). Microwave irradiation is more appropriate in some cases than other approaches (in the case of some breaks linking to cave or ground).

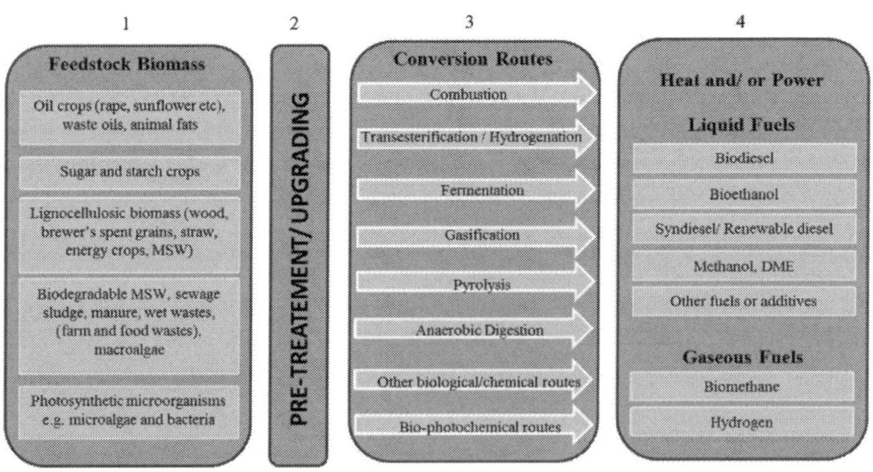

Figure 1.4 Current bioenergy methods which can be employed to various biomass feedstocks. *From: Bioenergy, I. (2009). Bioenergy—a sustainable and reliable energy source. International Energy Agency Bioenergy, Paris, France.*

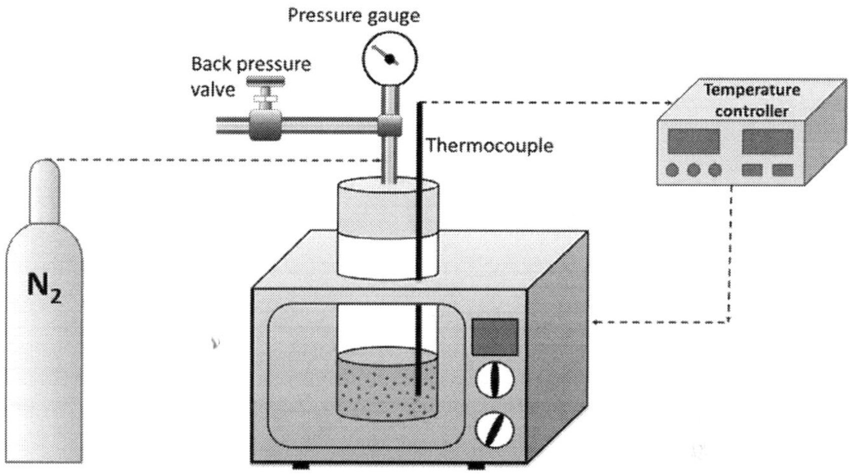

Figure 1.5 The schematic of a microwave-assisted heating system (Bach, Chen, Lin, Sheen, & Chang, 2017).

Microwave irradiation applying antennae can be a promising way for heating the reservoirs for coal seam degassing aims. This can be divided into two categories: (1) on the ground and (2) under the ground ways. For describing of application of microwave irradiation in the field of degassing coal seams, the conceptual plan is represented in Fig. 1.6. The systems consist several aspects as follows:

- A microwave heating system including a microwave source system, a voltage standing wave ratio (VSWR) sensing system, a line for microwave transmission, a control system, a system for sensing temperature, and an antenna.
- A CBM extraction system which consists of a floral and an extraction tube with a pump.
- A drainage hole place which contain a cylinder borehole part, a casing (for maintaining the mentioned borehole), and a sealing section (for preventing the methane leaking).

Nevertheless, the applying of microwave heating under the ground can possess more perilous compared to the ground processing. As illustrated in Fig. 1.6B, the microwave heating construction should be placed in the roadway. By the way, a methane-sensing section must place near the borehole to prevent methane from reaching to explosion levels. Furthermore, the microwave source should be located on the upwind (Hong, Lin, Zhu, & Li, 2016).

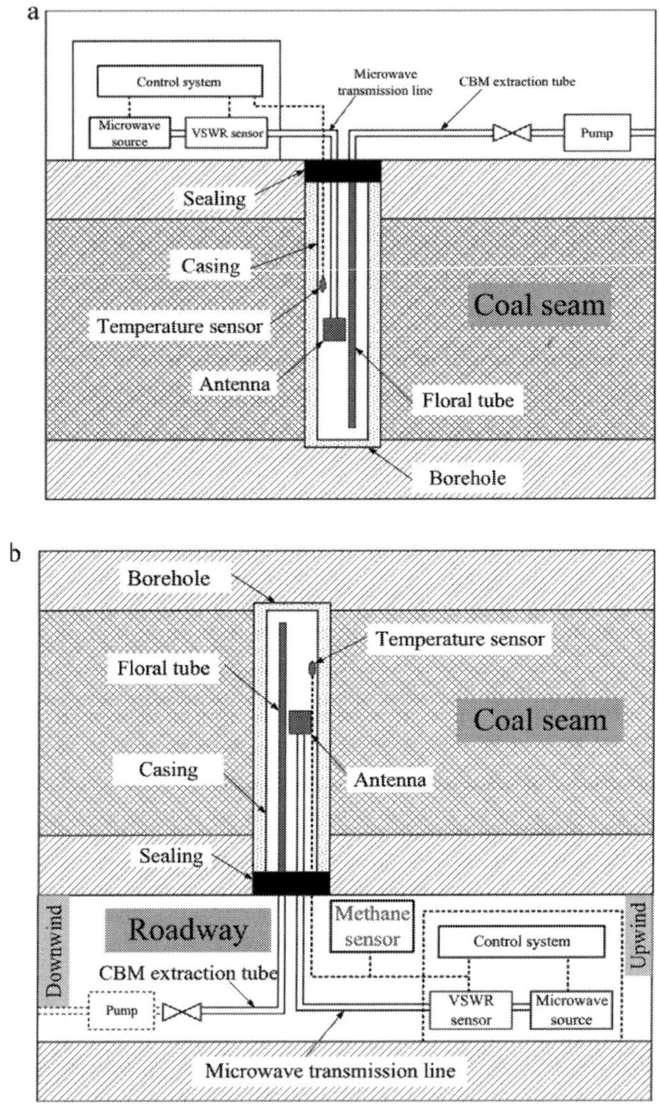

Figure 1.6 A schematic for microwave heating mechanism for the degassing coal seams: (A) ground drainage; and (B) underground drainage (Hong et al., 2016).

1.4 Synergistic effects

Process intensification via the combination of two individual methods or mechanisms can cause to synergism (magnified effect by both components). The term synergism can be expressed as a phenomenon creating from a combination of two techniques or methods, or agents that led to results higher than the sum of the values from the individual. In order to accomplish synergism, it is important that process intensification successfully address the following main factors (Huang, Wang, Shan, Zhu, & Qian, 2007; Van Gerven & Stankiewicz, 2009):

- Maximize the effectiveness degree of interactions (both intramolecular and intermolecular domain) via producing dynamic conditions for triggering kinetic regimes with better conversion and efficiency.
- To ensure a uniform gradientless mixing and heating process.
- Optimize inducing forces and maximize the surface to volume ratios to better both heat and mass transferring efficiencies.
- Raise the synergism from conventional or partial systems.

Recently, MW or US mediated organic synthesis is of great interest in focal point mainly owning to the superior effects obtained in shorter reaction periods and increased product yields. As these process intensification mechanisms have been well applied in various applications include chemistry and engineering fields, they are explored by biofuel industry more extensively because of their beneficial properties. These two nontraditional (new) irradiation processes have been employed in different fields as feedstock preparation, pretreatment efforts, extraction procedures, chemical conversion systems, and post-treatment steps in biofuel production protocols (Gude, Patil, Martinez-Guerra, Deng, & Nirmalakhandan, 2013). MWs deliver a compact created by the interaction between reaction materials and electromagnetic field which often lead to thermal amplification resulted in excellent results in chemical synthesis approaches (Fig. 1.7A). MWs possesses an ability of providing instant heat produced by three main mechanisms in the reaction conditions include ionic conduction, dipolar momentum and interfacial polarization (Fig. 1.7B).

In the other hand, US can produce acoustic cavitations as a result of the interactions between sound waves (Fig. 1.7C) and reaction materials lead to intensified mixing (Fig. 1.7D) that raise the transferring processes (both mass and heat domains) leading to better process efficiencies (Gude & Grant, 2013; Luo, Fang, & Smith, 2014). For instance, process intensification obtained from US can trigger mass transferring capability through gas and liquid moieties by fivefold magnitude. On the other hand, the liquid—solid mass transferring processes can be leveled by 20—25 fold (Fig. 1.8). In this way, the obtained reaction times drastically reduced using MW heating method by 1250 magnitude because of rapid enhancement of produced heat. The mentioned techniques can also amend the energy and material efficiencies owning to high levels of product conversion and efficiencies. Although MWs enable rapid heating of the materials, the mass transferring is often settled in these reactors (Martinez-Guerra & Gude, 2014a; Stankiewicz, 2006).

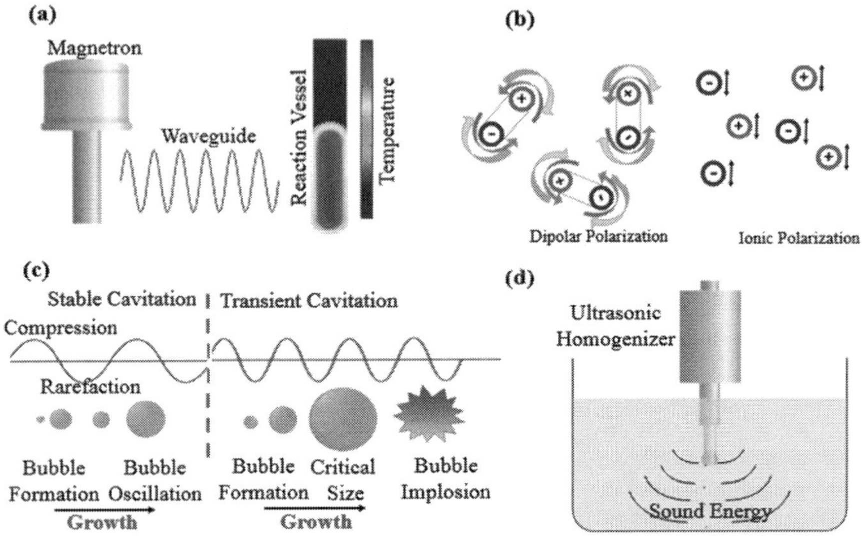

Figure 1.7 Different aspects of US and MI-S in chemical synthesis. *From: Gude, V. G. (2015). Synergism of microwaves and ultrasound for advanced biorefineries. Resource-Efficient Technologies, 1(2), 116—125.*

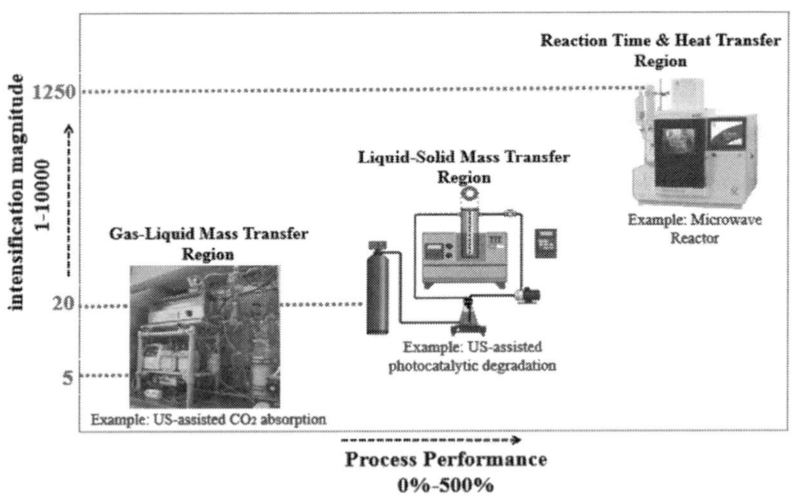

Figure 1.8 Process intensification compacts obtained from MWs and US. *From: Gude, V. G. (2015). Synergism of microwaves and ultrasound for advanced biorefineries. Resource-Efficient Technologies, 1(2), 116—125.*

In addition, they can interact with reaction compounds more intensely which lead to produce hot spot and thermal escaping. This phenomenon shows a mixing mechanism to be essential that lead to uniformly heating of the materials along with promoted mass transferring ability as a result of heating benefits of the MWs. On the other hand, US possesses the ability of triggering the transferring mechanisms (include heat and mass transferring) through the reaction substrate owning to the increased mixing resulted from the acoustic cavitation produced by microbubbles with air (Gude, 2015; Martinez-Guerra & Gude, 2014a). The mentioned reactions need an external heating source for enhancing the process kinetics.

Considering the mentioned viewpoints and restrictions for the intensification mechanisms in individual processing, it is of great interest to plan a hybrid strategy which combine both the mentioned nonconventional methods (for their heating and mixing effects). This may lead to increase the outcomes from process. This can address biocompatible chemistry (Fig. 1.9) which can efficiently apply chemicals, energy and compounds (Gude & Martinez-Guerra, 2015; Martinez-Guerra & Gude, 2014b). Excellent benefits obtained via the hybridized process intensifications might be economical friendly at large scales. It is substantial to note this hybrid as an ideal methodology for production various compounds.

As considering the above-mentioned states, the synergistic compact between microwaves and ultrasonic (US/MI-S) domain provide several desired advantages used in many aspects of science. Introduced a method

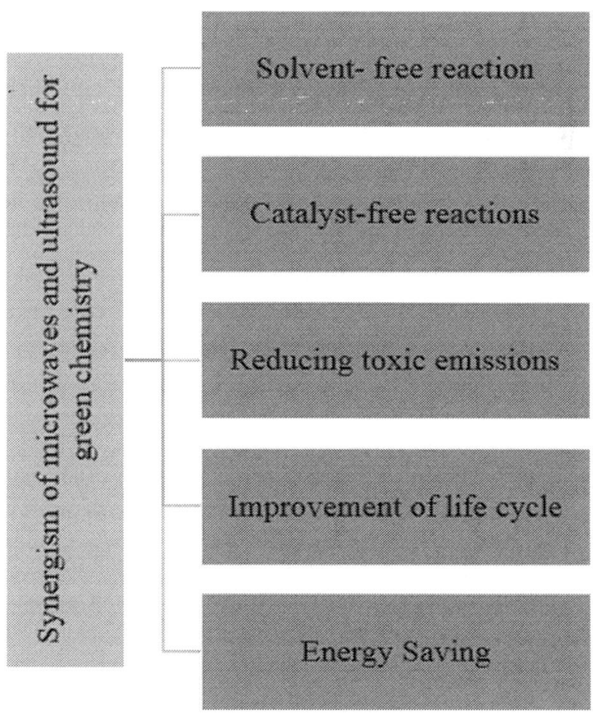

Figure 1.9 MW and US synergism based on green chemistry. *From: Gude, V. G. (2015). Synergism of microwaves and ultrasound for advanced biorefineries. Resource-Efficient Technologies, 1(2), 116–125.*

to optimize the biodegradation performance of dextranase enzyme employing the synergistic effects raised between ultrasound and microwave domains. In their study, the hydrolysis rate of dextran was investigated in three different conditions to obtain a view on US/MI-S. According to Fig. 1.10, the combination of US and MI-S increase the efficiency of the process (Bashari, Jin, Wang, & Zhan, 2016). The thermodynamic parameters from treatment of dextranase enzyme under different conditions of US, MI-S, US/MI-S, and conventional incubation conditions were briefly represented in Table 1.1.

Values are the mean ± standard deviation ($n = 3$). Means in the column with different letters are significantly different (*P* values of 0.05).

1.5 Separation processes using ultrasonic and microwave irradiation

1.5.1 Ultrasonic-assisted sorption processes

The main principal concepts of application of ultrasonic irradiation for enhancing mass transfer are described for extraction process, principal concepts and application of ultrasonic irradiation in sorption processes are relatively similar. In all extraction methods, the time needed to get the equilibrium state may be accelerated using different pathologies:

- Direct heating for increasing the mobility of material molecules which lead to increase the extraction efficiency;
- Microwave-assisted procedure;
- High speed mixing processes;
- Use of electrical discharge; and
- Ultrasonic-assisted extraction (UAE).

The first reported work on UAE is at 1950 which relates to the application of hops in the brewing of beer (Bamforth, 2017). The triggering force for UAE can be correlated to cavitation process which was first identified and characterized in 1895 (Mason, 2012). The mentioned study beside subsequent researches were all in the pathway of creating cavitation applying propellers driven by water for instance hydrodynamic cavitation way. Otherwise, another mode of cavitation has been proved which can be created using ultrasound waves. Similar to any sound wave, ultrasonic waves are speeded through an alternative mode of compression and rarefaction waves triggered in the media molecular components through which it goes passes. When the applied power is high enough, the rarefaction cycle can overstep the molecular attractive forces in the liquid media and led to produce cavitation bubbles (Fig. 1.11).

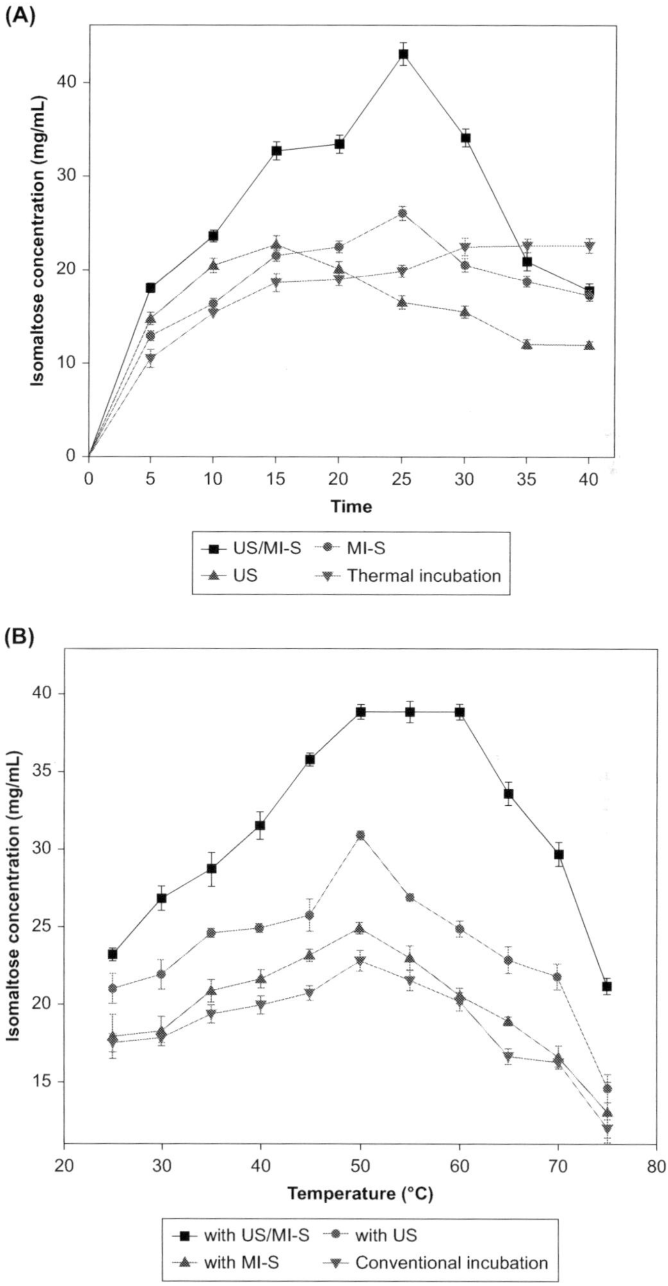

Figure 1.10 The comparison of the results obtained by US, MI-S, US/MI-S, and conventional conditions (Bashari et al., 2016).

Table 1.1 The effect of ultrasonic on the results, compared to the conventional method (Bashari et al., 2016).

Approach	ln (A)	Ea (kJmol^{-1})	ΔH (kJmol^{-1})	ΔS (kJmol^{-1})	ΔG (kJmol^{-1})
US/MI-S	20.01	45.36 ± 2.48	52.29 ± 2.38	−0.777 ± 0.0016	47.53 ± 1.39
US	22.17	60.24 ± 2.17	60.34 ± 1.76	−0.0895 ± 0.0011	54.85 ± 0.22
MI-S	23.09	62.77 ± 1.52	68.63 ± 1.46	−0.1018 ± 0.0086	62.38 ± 1.86
Conventional incubation	23.56	60.05 ± 2.96	70.47 ± 0.58	−0.1045 ± 0.0042	64.07 ± 1.18

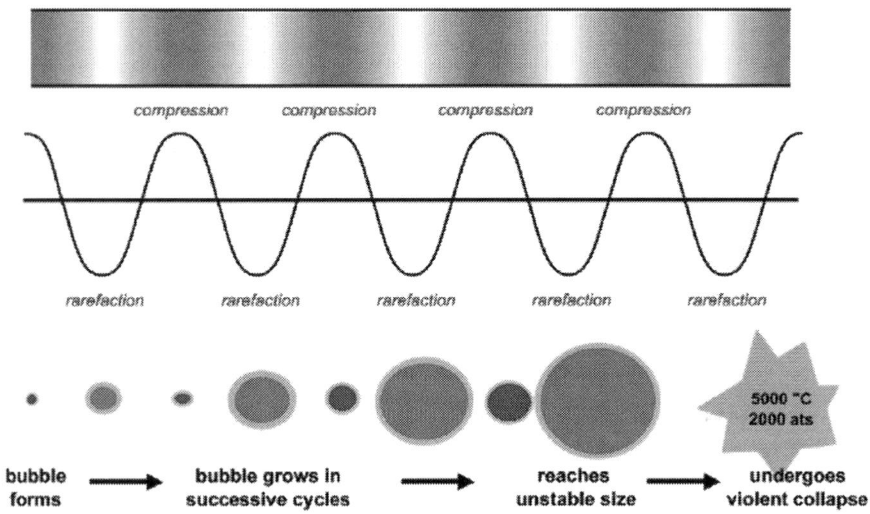

Figure 1.11 The creation mechanism for cavitation bubbles and the way of their collapse (Vinatoru, Mason, & Calinescu, 2017).

As mentioned above, US can be applied to increase mass transferring capability in media, which can be employed in continuous or batch type, or in two types of applying an ultrasonic bath or using a probe-based protocol for extraction purposes (Fig. 1.12A—D). There are two major forms of transducer commonly employed for ultrasonic-assisted usages: (1) piezoelectric and (2) magnetostrictive ways (Makino, Mossoba, & Riesz, 1983). Placement of the transducer has a critical role in detection of extraction efficiencies. It is of great important in the viewpoints of extraction efficiency, process intensification mechanism and energy losing cases. Transducers can be located on either side in the extraction vessel; therefore, US waves can transmit within the outer wall. The main benefit of the mentioned arrangement includes indirectly contraction of transducers with the sample (Fig. 1.12A and B). Multiple transducers have been employed in a flow-

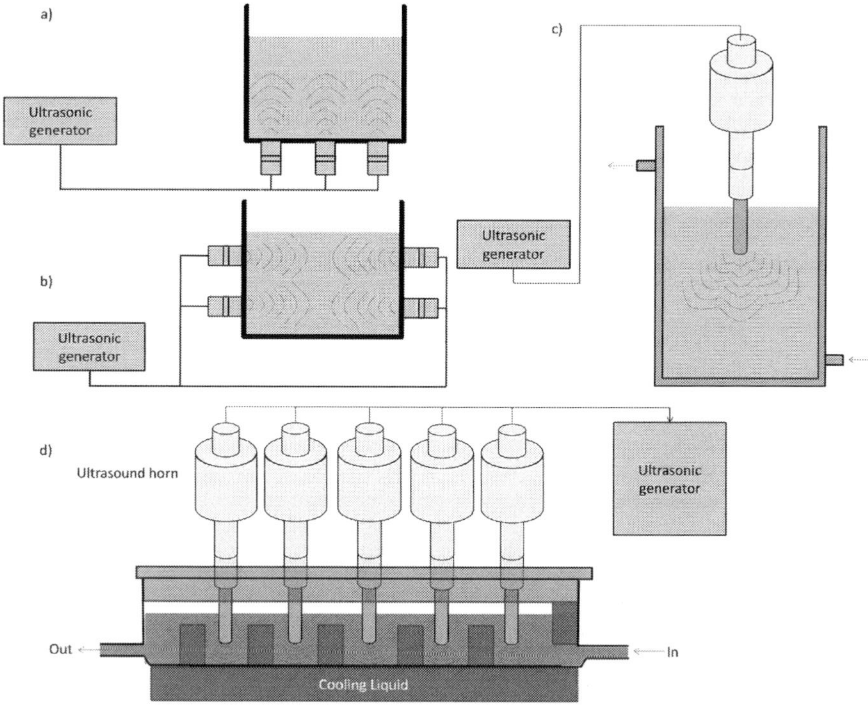

Figure 1.12 Different forms of ultrasonic-amplified extraction protocols: (A, B) illustrates ultrasonic bath models; (C) ultrasonic horn model; (D) horn-based extraction system using continuous ultrasonic waves (Fitzgerald et al., 1956).

through system in order to large-scale extraction aims and also to advance continuous extraction devices (Fig. 1.12D). in the case of direct connection of transducers with the sample which occurs in the presence of an appropriate solvent, the obtained extraction yield is increased, but acoustic energy losses are in minimized point.

The creating of a high-performance UAE system for large-scale extraction protocols needs process intensification strategy and the energy consumption to be reduced. This is due to reach safe, sustainable, economically friendly and greener strategies (Fitzgerald, Griffing, & Sullivan, 1956). Fig. 1.12D represents a continuous sono-reactor system acting at continuous way at higher acoustic power density (Vinatoru, Mason, & Calinescu, 2017). It should be mentioned that there are several kinetic models of ultrasound-assisted extraction has been reported up to date:

- Second-order model
- Weibull model

- Peleg's model
- Two-site kinetic model
- Unsteady state diffusion model based on the Fick's second law

Fig. 1.13 represents a system which always needs to be stirred the mixture to prevent vegetal mass agglomeration in sections of the container where no cavitation can be formed. Stirring is of great significant due to the fact that vegetal compounds can be floated or sank into the used solvent. Therefore, there will be an attraction to eschew ultrasonic active zones for some materials inside the bath in the absence of any stirring.

Fig. 1.14 represents different types of ultrasonic horn sonication. In this method several standard laboratory establishments can be exploited.

A larger construction which is more selectively designed for application with an ultrasonic horn mode is illustrated in Fig. 1.15 (Mason, 1999). The downward valve is applied for discharging reactor contain a large bore by which the solvent and vegetal mass let to pass through.

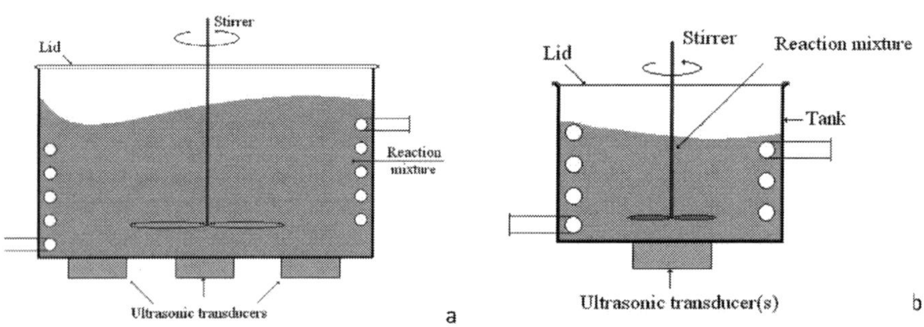

Figure 1.13 Direct sonication protocol: (A) large bath containing several transducers; (B) small bath containing single transducer (Vinatoru, Mason, & Calinescu, 2017).

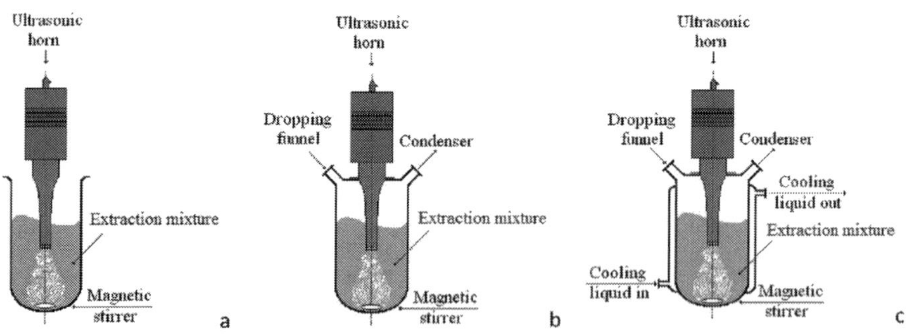

Figure 1.14 Possible models of ultrasonic horn sonication method: (A) open vessel mode; (B) close vessel mode (Vinatoru, Mason, & Calinescu, 2017).

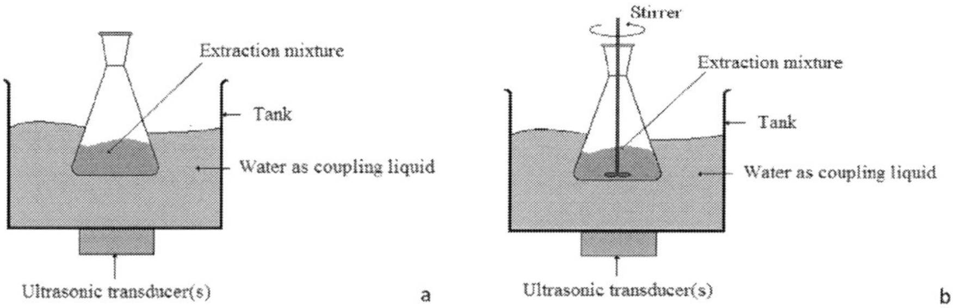

Figure 1.15 Extraction vessel locating into the tanker for indirect sonication purpose: (A) stirless extraction; (B) stirred assisted extraction (Vinatoru, Mason, & Calinescu, 2017).

Homogeneously and internally heating the whole volume of the sample is always termed as the major properties and benefits of microwave-assisted extraction (MAE). In other way, it also led to some benefits as follows:

- Good selectivity heating, and
- Declined installation footprint compared to the conventional approaches.

There are further benefits:

- Ease of production methodology (low from time consuming aspects), as much as about 75%–80%.
- Accelerated preparing process.
- Lower electrical energy consumption compared to traditional strategies between 25% and 50%.
- Low cost as a result of low production waste and losses.
- Reduction of the footprint as about 50%–90%.
- Shorter time for setting up and commissioning.
- Costly effective maintenance.

The application of the ultrasonic-assisted adsorption process for treatment of different pollutants is shown in Table 1.2 with respect to different influencing parameters and maximum adsorption capacity.

Ultrasonic-assisted technology has been evaluated to improve the mass transfer in the gas absorption process. High frequency ultrasonic irradiation has been applied for CO_2 capture using chemical solvent with its promising absorption enhancement and compact design (Tay, Lau, & Shariff, 2017).

Table 1.2 The application of the ultrasonic-assisted adsorption process for treatment of different pollutants (Mary Ealias & Saravanakumar, 2019).

Adsorbents	Pollutant	pH	Adsorbent Dosage (g)	Initial pollutant Concentration (mg/L)	Sonication Time (min)	Ultrasonic Frequency (kHz)	Ultrasonic power (W)	Adsorption capacity, Q_m (mg/g)	References
Fe_3O_4-MNPs-AC	Sunset yellow	4	0.04	15	6	40	130	76.37	Bagheri, Ghaedi, Asfaram, Bazrafshan, and Jannesar (2017)
	Methylene blue	6	0.03	15	6	40	130	78.76	
	Eosin B	5	0.032	25	4	40	130	102	
ZnS-NP-AC	Malachite green	6	0.025	22	5	0.06	130	51.55	Roosta, Ghaedi, and Yousefi (2015)
$Zn(OH)_2$-NP-AC	Malachite green	4.5	0.019	20	8.6	40	130	74.63	Bazrafshan, Hajati, and Ghaedi (2015)
AC-MOF-5	Fast green	7.06	0.02	5.68	2.55	40	130	21.23	Askari, Ghaedi, Dashtian, and Azghandi (2017)
	Eosin Y	7.06	0.02	7.59	2.55	40	130	20.24	
	Quinine yellow	7.06	0.02	5.04	2.55	40	130	18.62	
γ-Fe_2O_3-NP-AC	Reactive red 223	11	0.02	25	0.4	>20	–	57.58	Saad and Tahir (2017)
	Malachite green	11	0.016	11	6.80	>20	–	29.32	
AlNs	Congo red	3.5–10.5	0.16	50	10	80	70	121.95	Ealias and Saravanakumar (2018)
	Crystal violet	9.8	0.12	100	10	80	70	105.26	
NOAC	Basic red 46	9	0.03	150	8	40	130	1111	Khaligh, Mousavi, Rashidi, and Shirkhanloo (2018)
	Basic yellow 13	9	0.03	150	8	40	130	1250	

MWCNTs	Methyl orange	1	0.025	15	3.4	40	130	53.76	Ghaedi et al. (2015)
ZnS-NP-AC	Brilliant green	5	0.0175	12	3.78	40	130	258.7	Azqhandi, Ghaedi, Yousefi, and Jamshidi (2017)
ZnO-NRs-AC	Bromocresol green	6	0.02	9	4	0.06	130	57.80	Ansari, Ghaedi, Taghdiri, and Asfaram, (2016)
	Eosin Y	6	0.02	10	4	0.06	130	61.73	
ZnO-NP-AC(AS)	Basic blue 41	6	0.03	9	19	50/60	–	0.603	Asghari, Hemati, Ghaedi, Rajabi, and Mirtamizdoust (2014)
	Basic red 46	6	0.03	14	19	50/60	–	0.442	
ZnO-NR-AC	Malachite green	6	0.035	12	4	40	130	20	Ghaedi et al. (2016)

The enhancement for the chemical absorption using ultrasonic irradiation is believed to be dominated by the fountain formation and also the convective dynamic (Fig. 1.16).

The volumetric absorption coefficient by using ultrasonic irradiation was nearly three times bigger than the stirring method, as shown in Table 1.3. In comparison with MEA, the less improvement can be related to this fact that the CO_2 reaction mechanism with a slow kinetic solvent, such as MDEA, is totally different, which is not greatly affected by the physical effects created by the ultrasonic irradiation. However, in this case, the presence of the ultrasonic chemical effect might be more effective and dominant. It is believed that the sonochemical effect is able to improve the kinetic rate of the chemical reaction by changing the reaction pathway.

The application of ultrasonic-assisted absorption process for treatment of different gas species is shown in Table 1.4.

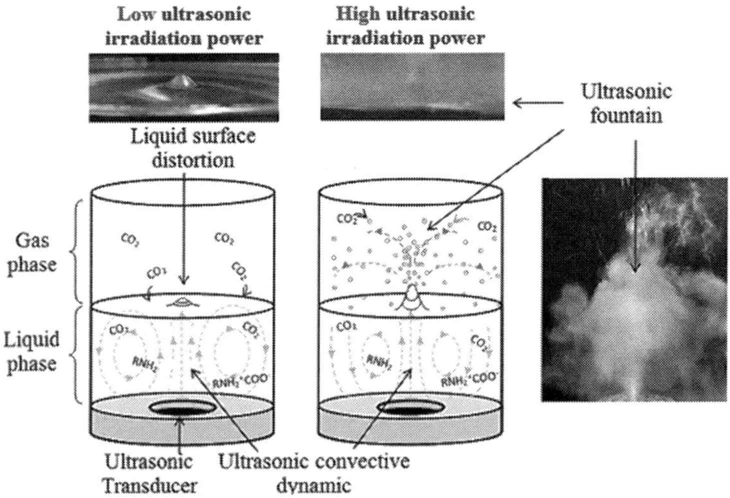

Figure 1.16 The fountain formation and the convective dynamic. Low ultrasonic irradiation power (left-side) and high ultrasonic irradiation power (right-side) (Tay et al., 2017).

Table 1.3 Comparison of absorption rates between ultrasonic and stirring techniques (Shokrollahi, Lau, & Tay, 2020).

Applied method	Specification	P_{CO2} (kPa)	T(K)	Absorption rate * 10^{-3} mol/s	Volumetric absorption coefficient * 10^{-5}mol/pa .m^{-3}.s^{-1}
Stirring	500 rpm	1100	333	0.047	0.086
Ultrasonic	18 W	1100	333	0.667	0.248

Table 1.4 Ultrasonic-assisted mass transfer for the absorption process (Shokrollahi et al., 2020).

Operation mode	Gas	Absorbent	Ultrasonic effect	Ultrasonic power	Ultrasonic frequency	Findings	References
Semibatch	O_2	Water	Physical	600	22 kHz	Mass transfer coefficient increases with an increase in the power	Kumar, Gogate, Pandit, Wilhelm, and Delmas (2005)
Semibatch	O_3	Water	Physical	260	20 kHz	The presence of u ltrasound-enhanced ozone mass transfer rate	Zhang, Duan, and Zhang (2007)
Batch	N_2	Water	Physical	62.6	20 kHz	Ultrasound greatly improves gas liquid mass transfer	Laugier, Andriantsiferana, Wilhelm, and Delmas (2008)
Semibatch	NO	Water	Chemical	110	20 kHz	Both physical and chemical effects are dominant in the mass transfer enhancement	Adewuyi and Khan (2012)
Batch	CO_2	Water	Physical	18	1.7 MHz	The streaming force and fountain formation are the dominant factors for the absorption enhancement	Ma et al. (2008)

1.5.2 Microwave-assisted sorption processes

Microwave-assisted sample pretreatment provide an opportunity to hybrid the fast heating capability obtained from microwave energy with the benefits from sealed digestion containers. Microwave methods have several advantages include: fast, easy to automation and controlling compared to the conventional techniques (Li, Qu, Yang, Chang, & Xu, 2016).

The use of microwaves is well-recognized in the food processing sector and in other process industries. There have been attempts to use microwaves in catalytic microreactors, following on from success in using this energy form in organic synthesis, but results to date have not been positive. The authors of a research paper (Cecilia, Kunz, & Turek, 2007) have suggested that modifying the properties of composite material in which the palladium catalyst was loaded could lead to a positive outcome. Fig. 1.17 represents a schematic view of competition the heat generation pathways involve in the conventional and microwave-assisted methodologies.

The effective heating of different materials induced by microwave radiation mechanism is mainly rely on the physical characteristics of the applied materials. These characteristics define the interactions between microwave radiations and the employed materials. Microwaves cannot be absorbed by transparent compounds lead them to passes through these materials with no losses in the energy and so there is no heating incidence to be occurred. The conductor compounds reflect the microwaves back and lead to plasma formation as a result cause to surficial heating phenomena. But materials can willingly absorb microwaves, from which the irradiation converts into heat (Singh et al., 2014). The application of the microwave-assisted adsorption process for treatment of different pollutants is shown in Table 1.5.

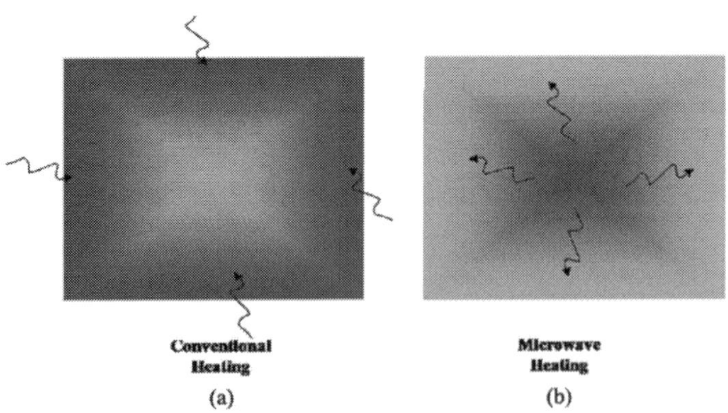

Figure 1.17 Heat generation process in (A) conventional heating models and (B) microwave heating models (Singh, Gupta, Jain, & Sharma, 2014).

Table 1.5 Application of MW-induced catalytic degradation (MICD) and MW-induced catalytic oxidation (MICO) processes (Wang & Wang, 2016).

	Experiment conditions						
Pollutant	MW time (min) and power (W)[a]	Catalyst (C)-oxidant (O) names and dosage (g/L)[b]	Initial pH	Others[c]	Pollutant, COD and/or TOC removal rate and energy efficiency	Additional information	References
Carbofuran	30 (time) 750 (Power)	GAC (C)-0.8	8	WW Con. = 100 mg/L; WW Vol. = 250 mL; Stirring speed = 120 rpm; T = 80°C.	≈100.0% (carbofuran) 72% (COD) 301.3 μmol/kW·h	More than 99% of removal rate (carbofuran) was achieved in all the three cycles of GAC; the destruction of carbonate group and oxidation of furan ring were the main reason for degradation.	Remya and Lin (2015)
Amoxicillin	7 480	GAC (C)-6	9	COD = 1334 mg/L after 10 times dilution; WW Vol. = 50 mL.	COD = 1334 mg/L after 10 times dilution; WW Vol. = 50 mL.	The increase of activated carbon amount and MW irradiation power are both beneficial for the effective removal of Amoxicillin.	Feng, Wang, Jia, and Ning (2013)

(Continued)

Table 1.5 Application of MW-induced catalytic degradation (MICD) and MW-induced catalytic oxidation (MICO) processes (Wang & Wang, 2016).—cont'd

	Experiment conditions						
Pollutant	**MW time (min) and power (W)[a]**	**Catalyst (C)-oxidant (O) names and dosage (g/L)[b]**	**Initial pH**	**Others[c]**	**Pollutant, COD and/or TOC removal rate and energy efficiency**	**Additional information**	**References**
p-nitrophenol (p-NP)	8 500	GAC (C)-6	N/A	WW Con. = 200 mg/L; WW Vol. = 50 mL.	99.0% (p-NP) 1067.5 μmol/kW·h	p-NP initial concentration, MW power, MW irradiation time and GAC dosage are the significant factors in the degradation of p-NP.	Yuan, Fu, and Luo (2013)
Methyl Orange (MO)	8 539	GAC (C)–N/A	N/A	WW Con. = 20 mg/L; WW Vol. = 50 mL.	98.3% (MO) 41.8 μmol/kW·h	The activity of GAC can be effectively recovered by MW radiation and 74.1% of its initial activity remains after six cycles; the performance decrease of GAC after repetitive use is largely due to the adsorption of some intermediate products on the surface of GAC.	Xu, Cheng, Zhang, and Song (2014)

Phenol	5 750	Nano-Ni/ZnO (C)-0.0667	7	WW Con. = 100 mg/L; WW Vol. = 30 mL; T = 70°C.	8.0% (phenol) 499.8 μmol/kW·h	MW oven was operated at 750 W that heated the sample solution from room temperature to 70°C and then was maintained for 2 min; the addition of H_2O_2 led to the increase of catalyst leaching.	Ahmed, Jibril, Danwittayakul, and Dutta (2014)
Petroleum Refinery WW	30 660	GAC (C)-50 O_2 was supplied with 0.8 Mpa.	N/A	COD = 5500 mg/L; WW Vol. = 150 mL; T = 150°C; *P* = .8 MPa.	91.0% (COD) 2275.0 mg COD/kW·h	BOD5/COD ratio was increased from 0.04 to 0.47 within 30 min in the CWAO system.	Sun, Zhang, and Quan (2008)
Remazol Golden Yellow	1.5 400	CuOn-La_2O_3/ r-Al_2O_3 (C)- 70 ClO_2 (O)-0.08	7	WW Con. = 200 mg/L; WW Vol. = 100 mL; T = 88°C.	94.0% (decoloriztion) 67.9% (TOC) 2244.2 μmol/kW·h	Variation of the color removal rate was not linearly related to temperatures.	Bi, Wang, Jiao, and Cao (2009)

(*Continued*)

Table 1.5 Application of MW-induced catalytic degradation (MICD) and MW-induced catalytic oxidation (MICO) processes (Wang & Wang, 2016).—cont'd

	Experiment conditions						
Pollutant	**MW time (min) and power (W)[a]**	**Catalyst (C)-oxidant (O) names and dosage (g/L)[b]**	**Initial pH**	**Others[c]**	**Pollutant, COD and/or TOC removal rate and energy efficiency**	**Additional information**	**References**
p-NP	300 (hydraulic retention Time) 400	Cu/GAC (C)–N/A Air flow = 120 mL/min (air and p-NP solution were mixed and pumped Into the fixed bed reactor; Continuous flow mode)	N/A	WW Con. = 1734 mg/L; p-NP solution flow rate = 6.4 mL/min; Ambient pressure; 150–160°C in outflow.	91.8% (p-NP) 88% (TOC) 10,985.1 μmol/kW·h	GAC structure was enhanced by the load of Cu. A higher catalytic activity and better stability of Cu/GAC catalyst than GAC wereobserved.	Bo, Zhang, Quan, and Zhao (2008)
Phenol	5 50	CuOn–La_2O_3/r-Al_2O_3(C)–50ClO_2(O)-0.08	3–7	WW Con. = 100 mg/L; WW Vol. = 100 mL; T = 47°C	91.0% (Phenol)50.4% (TOC) 23,206.9 Lmol/kW·h	MW assisted ClO_2catalytic oxidationprocess using CuOn–La_2O_3/r-Al_2O_3 ascatalyst can degrade contaminantsin short reaction time and with lowoxidant dosage, wide pH ran	Lv et al. (2009)

[a]MW Frequency is 2450 MHz.
[b]No. 1–5 are MICD process; No. 6–8 are MICO process.
[c]Con. is the abbreviation for concentration; Vol. is the abbreviation for volume; T means temperature.

Table 1.6 Microwave ODS efficiency using different oxidizing agents (Shang et al., 2013).

			ODS efficiency (%)	
Oil type	**Desulfurizers**	**Dosage (%)**	**With MW**	**Without MW**
Crude oil (Lu, Yang, Xu, & Gao, 2009)	DCP	1	22.1	12.6
	BPO	1	22.1	21.8
	BBPV	1	34.5	28.5
	BMPC	1	43.3	24.3
	H_2O_2	5	16.9	12.5
	CH_3COOOH	5	34.7	21.8
	HCOOOH	5	33.3	18.6
DBT	BBPV	0.5/4	48.0/64.2	–
	BPMC	0.5/4	4.7/38.0	–
BT	BBPV	0.5/4	19.8/53.7	–
	BPMC	0.5/4	19.8/46.6	–
	HCOOH	5/20	29.4/60.8	–
	CH_3COOOH	5/20	40.3/60.8	–

The microwave field can be used for enhancement of the gas absorption efficiency. Direct desulfurization of fuel petroleum oils has been studied under microwave irradiation which results indicated that good sulfur removal efficiency from oils may be achieved under microwave irradiation (Shang, Zhang, Du, & Liu, 2013). Table 1.6 summarizes the oxidative desulfurization (ODS) efficiencies under microwave irradiation using different oxidizing agents.

Along with the development of microwave aided desulfurization research, as well as the requirements of the industrialization, microwave device selection and development have become an important topic, its high efficiency, stability, safety, processing power, and scale are all the well worth study directions (Shang et al., 2013).

1.6 Economics and cost evaluations

For justifying the process economics in full-scale applications, a cost-benefit analysis is essential comprising treatment cost (capital, operation, and maintenance) as well as achieved paybacks (biogas recovery, reducing the ultimate solids content) of ultrasonic integrated systems. The major costs for an already working conventional system include sludge disposal and transportation. Thus, the incremented biogas and decremented disposal and transportation costs can be considered as revenues and upgraded situation savings. The economic advantages of the ultrasonic treatment has been assessed for sludge treatment and consequent anaerobic digestions (Appels, Baeyens, Degrève, & Dewil, 2008) which findings reveals that the investment cost of about $52,400 for ultrasonic treatment as the maximum of 20 kW to treat sludge from a wastewater treatment plant (WWTP) of 10,000 population equivalent (p.e.). The costs of operation and maintenance are minimal, however, the ultrasound probes require replacement each 1.5–2 years. Moreover, a payback period of fewer than 2 years is beneficial. Considering the fast sludge degradation with enhanced biogas production (+30%) and solids destruction (+30%), the digesters' dimensions can be decreased for a considered load. Therefore, the impact of high capital requirements is reduced. Dhar et al. indicated that ultrasound pretreatment is economically feasible only at the definite energy input of 1000 kJ/kg total suspended solid (TSS), by net savings of $54/ton dry solid (Dhar, Nakhla, & Ray, 2012).

1.7 Industrial upscaling

Ultrasound technology studies for environmental engineering have been initiated in almost the half 1990s mostly in Germany. It was aimed to overwhelm the rate limiting hydrolysis stage of the anaerobic sludge stabilization via ultrasound disintegrating sludge cells and afterward intensifying the anaerobic degradation procedure. Finally, less residual and more biogas sludge was the resultant. Followed by the turn of the millennium, first full-scale ultrasound installations were emergent in Germany. Today further attention is shifted to optimize the treatment plants operation to decrease the energy consumption and other expenses in an economic sense, as well as for ecological aspects like the reduction of greenhouse gas emissions (Neis, 2015).

The different engineering industries have some problems for surface degradation of the components, mainly owning to the corrosion and wear issues (Dawei, Zhang, Xie, & Yazhi, 2009) which can lead to degradation of component. To overcome the problem, many surface modification methods like carburizing, nitriding, cyaniding, have been reported.

Recently, microwaves application for claddings purposes has been investigated by researchers (Chen et al., 2016). It has been uncovered that cladding process by applying microwave heating lead to lower defects such as low solidification cracks and declined porosity which can be related to increased mechanical characteristics. The domestic microwave oven can be applied for low- and high temperature applications in food stuffs heating. The adhering two different metals with low levels of defects is of great vital importance for manufacturing industries. The adhering bulk metallic compounds applying microwaves has recently possessed many researchers (Barnier, Ouvrard, Robin, & Morel, 2014; Bocos, Fernández-Costas, Pazos, & Sanromán, 2015). To study in this domain. The experimental study by Srinath et al. (Abramovitch, ChangQing, Hicks, & Sinard, 2003) Probed a new method for metallurgical adhering of high thermal conductive compounds (such as copper) applying microwave heating method. It was mentioned that adhering of a metal like copper employing traditional approaches possesses several issues because of high conductivity nature. Also, remediation/detoxification process is of microwaves applications in industry.

MW irradiation has been adopted widely as a cost-effective method as an alternative approach for several heating technologies in several domains of industry such as mineral processing and metal extractions, drying methodologies, cement and concrete processing, food field and oil processing (Falciglia et al., 2018). Using the above-mentioned advantages, microwaves can be beneficial in extraction of several materials such as organic pollutants (Wang, Ding, & Ren, 2016), phenolic compounds (Flórez, Conde, & Domínguez, 2015; Ganzler & Salgo, 1987), pesticides (Wu et al., 2016), and so forth. Applying solvent free methods for extraction and microextraction of compounds is of great interest regarding the costs and hazards of the toxic organic solvents. In this regard, microwaves provide a domain for this purpose. Several researches have been reported use of microwaves for extraction of materials from aromatic herbs (Filly et al., 2014) olive tree leaves (Şahin et al., 2017). It is used for biodiesel production from microalgae (Ma et al., 2014).

Microwaves are also applied for chemical processing and producing and synthesizing materials. In this way, the energy produced by microwaves can be a triggering and inducing the reactions (Dallinger & Kappe, 2007;

Nüchter, Ondruschka, Bonrath, & Gum, 2004). By emerging nanotechnology and requiring new method for synthesis of nanoparticles in safe and nontoxic way the microwaves have taken many attentions for its above-mentioned characteristics (Baghbanzadeh, Carbone, Cozzoli, & Kappe, 2011; Chen, Lee, & Liu, 2002; Pal, Shah, & Devi, 2009). In such the way, several nanomaterials have been produced such as metallic nanoparticles (Tsuji, Hashimoto, Nishizawa, Kubokawa, & Tsuji, 2005) silver nanoparticles (Hu, Wang, Wang, Zhang, & Yu, 2008), gold nanoparticles (Wang, Yu, Xiao, & Li, 2014), and so forth. Application of microwaves for synthesis of silver nanomaterials, as an important nanostructure, is illustrated in Fig. 1.18. As can be seen, in opposite of conventional heating, applying microwaves irradiation increases the temperature of the whole reaction media simultaneously.

The ultrasonic domain has been possessed several industrial aspects such as meat quality (Alarcon-Rojo, Carrillo-Lopez, Reyes-Villagrana, Huerta-Jiménez, & Garcia-Galicia, 2018), food drying enhancement (Musielak, Mierzwa, & Kroehnke, 2016; Paniwnyk, 2017) and so forth. The ultrasonic waves applied in many filed and domains. Their role in adsorption and intensification of the extraction and adsorption processes can be very useful in many fields such as water purification and extraction and removal of pollutants from the water or organic solutions. In this regard, many attempts to apply them in the mentioned aspects. For instance, applying in removal of toxic organic dyes (Dastkhoon et al., 2015; Jamshidi, Ghaedi, Dashtian, Hajati, & Bazrafshan, 2016). In this way, the ultrasound waves were exerted to a solution of pollutants and adsorbents to intensify their adsorption properties. Fig. 1.19 shows the application and effect the ultrasound on the removal of organic materials from the water (Midathana & Moholkar, 2009).

Considering the before mentioned advantages for both ultrasonic and microwaves, many attempts have been done for their application in various industrial and experimental fields. One of the advances in this regard is the applications in food processing domains, where ultrasound and microwaves lead to increase the quality of the produced foods (Knorr, Zenker, Heinz, & Lee, 2004). Also, their application in the separation of the different materials mentioned in previous sections is of their analytical usages which help scientist to improve their performances. Another field is

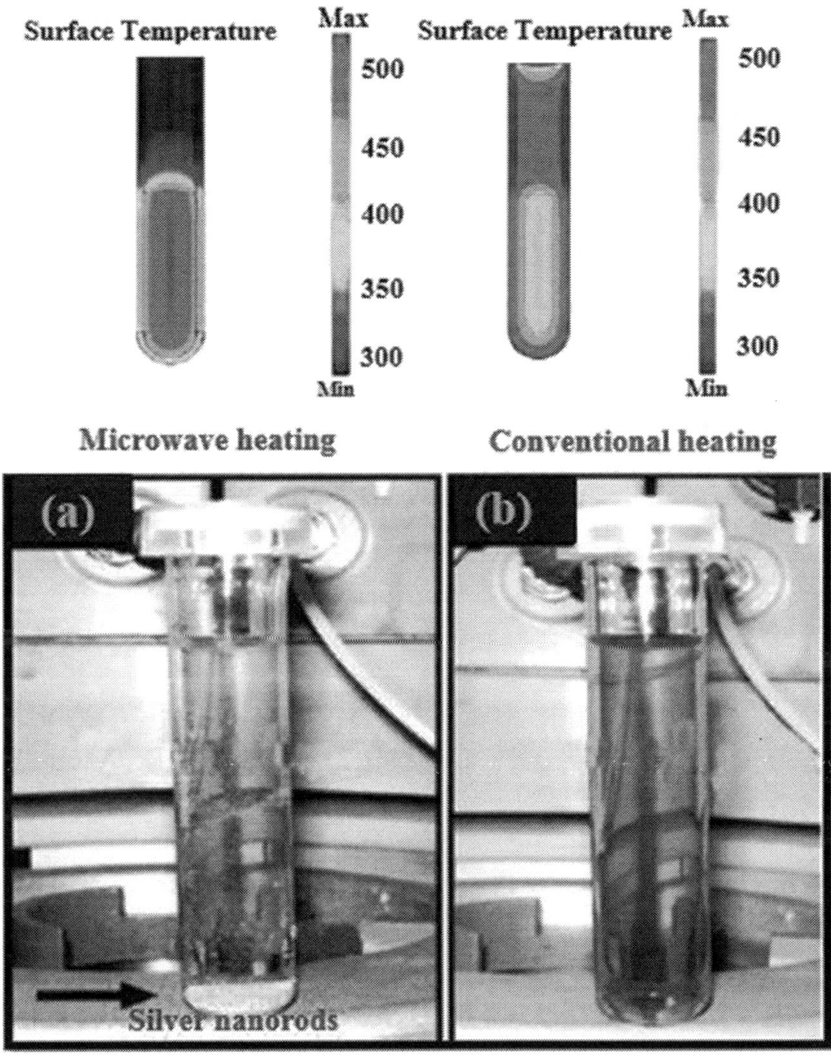

Figure 1.18 Temperature profiles for two different heating method for producing silver nanoparticles (Nadagouda & Varma, 2008).

wastewater treatment issue. This can be related to the molecular level heating capabilities of the both proposed methods (Khanal, Grewell, Sung, & Van Leeuwen, 2007). Also, many researches have been reported for the application in biomedical and point of care field (Kim et al., 2012; Mason, Korostynska, Ortoneda-Pedrola, Shaw, & Al-Shamma'a, 2013).

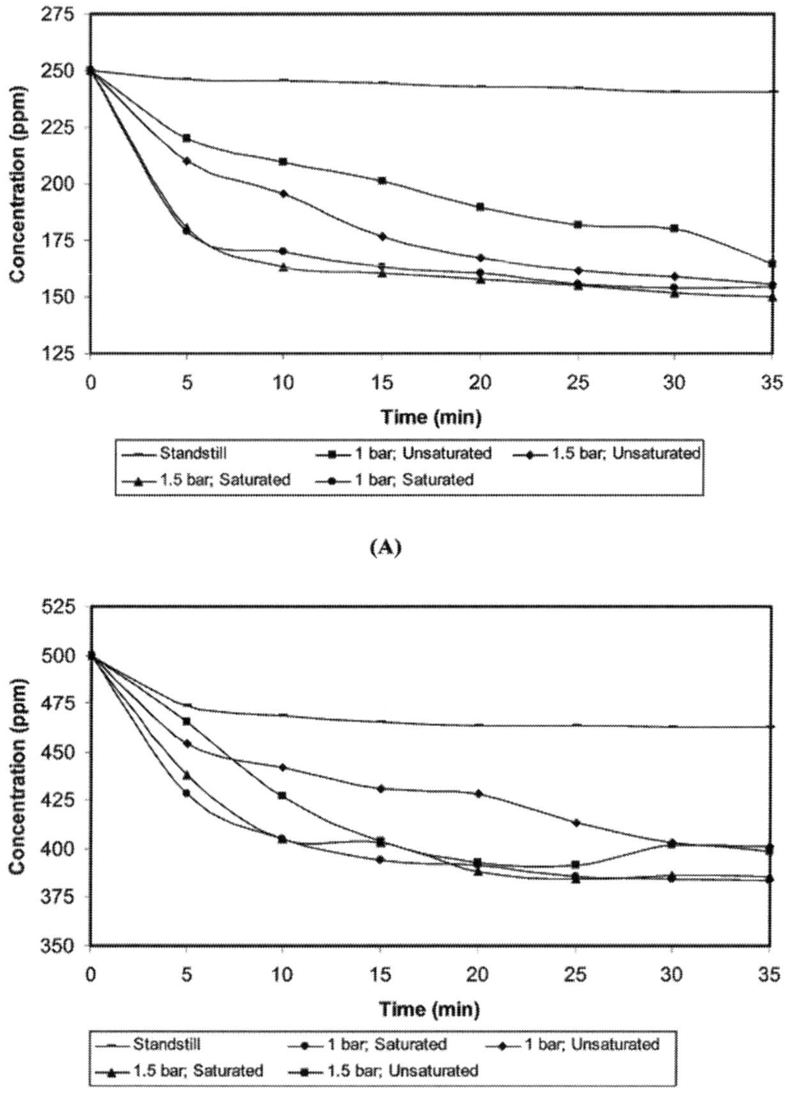

Figure 1.19 Trends in ultrasound-assisted protocol for adsorption of phenol on activated charcoal substrate: (A) initial concentration of 250 ppm, and (B) initial concentration of 500 ppm (Midathana & Moholkar, 2009).

1.8 Future perspectives

Although there are several important benefits of the MAE and ultrasound waves employed in the extraction of valuable compounds, the number of industrial-scale employments remain very small. On the other hand, it is of great interest to apply ultrasound waves in biomass processing. The industrial applications of MW are still restricted by the lack of many information and perceptions, which indicate some challenges associated with the application of MW heating method from which the MW is scarcely employed at the full-scale. Especially, full-scale research is still very restricted because of the complexity of designing MW heating devices.

Methodically, ultrasonic cavitation was investigated over the last years. The technology has been found to be an advanced, however, a small-scale and effective technique is still a challenge for wastewater and water treatment and commercial application owing to the higher energy demands of the process. Sonochemistry is of the most modern techniques for handling harmful water contaminants. The advantages of ultrasonic technology are as follows (Fetyan & Salem Attia, 2020):

- Lower capital cost and simple modular model
- Easy upgrading the conventional treatment units
- Lack of ultrasound tools additives and produced byproducts
- Lack of anticipated environmental concerns
- Higher performance of various pathogens from water and wastewater inaction based on the microorganism form.
- Oxidation of natural organic materials
- The increased performance by incrementing suspended solids or turbidity within the chemical degradation processes
- No traditional disinfection byproducts (such as CDMs)
- High synergy effect or efficiency enhancements

Designing properties of most devices established following empirical data including theoretical criteria should be established to rationalize these systems design and their larger scale application. Generally, this technology has the disadvantages as follows (Fetyan & Salem Attia, 2020):

- Developing development criteria
- Rising water turbidity
- Consumption of Energy

- Conservation or substitution of the ultrasound probe
- No accessible disinfection power

Most of operating costs are related to the keeping or replacing tools continuously damaged by ultrasonic activity.

1.9 Conclusion

In this section we have discussed ultrasound and microwaves in several aspects. These domains of the science have been represented many applications such as extraction of compounds, adhering the two pieces of the metals, in food industries and for modification purposes of the compounds. It can be obvious that applying these two methods can possess several benefits such as ease to sue, cost effective, and overcoming the many limitations of the conventional methods. High intensity ultrasound waves create acoustic cavitation into a liquid provide physical forces that are regarded as the major mechanism responsible for the produced alterations in exposed compounds.

References

Abd Malek, M. N. F., Hussin, N. M., Embong, N. H., Bhuyar, P., Rahim, M. H. A., Govindan, N., et al. (2020). Ultrasonication: A process intensification tool for methyl ester synthesis: A mini review. *Biomass Conversion and Biorefinery*, 1–11.

Abramovitch, R. A., ChangQing, L., Hicks, E., & Sinard, J. (2003). In situ remediation of soils contaminated with toxic metal ions using microwave energy. *Chemosphere, 53*(9), 1077–1085.

Adewuyi, Y. G., & Khan, N. E. (2012). Modeling the ultrasonic cavitation-enhanced removal of nitrogen oxide in a bubble column reactor. *AIChE Journal, 58*(8), 2397–2411.

Ahmed, A. B., Jibril, B., Danwittayakul, S., & Dutta, J. (2014). Microwave-enhanced degradation of phenol over Ni-loaded ZnO nanorods catalyst. *Applied Catalysis B: Environmental, 156*, 456–465.

Alarcon-Rojo, A. D., Carrillo-Lopez, L. M., Reyes-Villagrana, R., Huerta-Jiménez, M., & Garcia-Galicia, I. A. (2018). Ultrasound and meat quality: A review. *Ultrasonics Sonochemistry, 55*, 369–382.

Ansari, F., Ghaedi, M., Taghdiri, M., & Asfaram, A. (2016). Application of ZnO nanorods loaded on activated carbon for ultrasonic assisted dyes removal: Experimental design and derivative spectrophotometry method. *Ultrasonics Sonochemistry, 33*, 197–209.

Appels, L., Baeyens, J., Degrève, J., & Dewil, R. (2008). Principles and potential of the anaerobic digestion of waste-activated sludge. *Progress in Energy and Combustion Science, 34*(6), 755–781.

Asghari, A., Hemati, M., Ghaedi, M., Rajabi, M., & Mirtamizdoust, B. (2014). Ultrasonic assisted adsorption of basic dyes from binary component systems onto ZnO nanoparticles loaded on activated carbon derived from almond shell: Optimization by central composite design. *Journal of Nanostructures, 4*(1), 17–30.

Askari, H., Ghaedi, M., Dashtian, K., & Azghandi, M. H. A. (2017). Rapid and high-capacity ultrasonic assisted adsorption of ternary toxic anionic dyes onto MOF-5-activated carbon: Artificial neural networks, partial least squares, desirability function and isotherm and kinetic study. *Ultrasonics Sonochemistry, 37*, 71–82.

Azqhandi, M. A., Ghaedi, M., Yousefi, F., & Jamshidi, M. (2017). Application of random forest, radial basis function neural networks and central composite design for modeling and/or optimization of the ultrasonic assisted adsorption of brilliant green on ZnS-NP-AC. *Journal of Colloid and Interface Science, 505*, 278–292.

Bach, Q.-V., Chen, W.-H., Lin, S.-C., Sheen, H.-K., & Chang, J.-S. (2017). Wet torrefaction of microalga *Chlorella vulgaris* ESP-31 with microwave-assisted heating. *Energy Conversion and Management, 141*, 163–170. https://doi.org/10.1016/j.enconman.2016.07.035

Bae, J.-S., & Bhatia, S. K. (2006). High-pressure adsorption of methane and carbon dioxide on coal. *Energy and Fuels, 20*(6), 2599–2607.

Baghbanzadeh, M., Carbone, L., Cozzoli, P. D., & Kappe, C. O. (2011). Microwave-assisted synthesis of colloidal inorganic nanocrystals. *Angewandte Chemie International Edition, 50*(48), 11312–11359.

Bagheri, A. R., Ghaedi, M., Asfaram, A., Bazrafshan, A. A., & Jannesar, R. (2017). Comparative study on ultrasonic assisted adsorption of dyes from single system onto Fe_3O_4 magnetite nanoparticles loaded on activated carbon: Experimental design methodology. *Ultrasonics Sonochemistry, 34*, 294–304.

Bamforth, C. (2017). Progress in brewing science and beer production. *Annual Review of Chemical and Biomolecular Engineering, 8*, 161–176.

Barnier, C., Ouvrard, S., Robin, C., & Morel, J. L. (2014). Desorption kinetics of PAHs from aged industrial soils for availability assessment. *The Science of the Total Environment, 470*, 639–645.

Bashari, M., Jin, Z., Wang, J., & Zhan, X. (2016). A novel technique to improve the biodegradation efficiency of dextranase enzyme using the synergistic effects of ultrasound combined with microwave shock. *Innovative Food Science and Emerging Technologies, 35*, 125–132. https://doi.org/10.1016/j.ifset.2016.04.007

Bauen, A., Berndes, G., Junginger, M., Londo, M., Vuille, F., Ball, R., et al. (2009). *Bioenergy: A sustainable and reliable energy source. A review of status and prospects.*

Bazrafshan, A., Hajati, S., & Ghaedi, M. (2015). Synthesis of regenerable $Zn(OH)_2$ nanoparticle-loaded activated carbon for the ultrasound-assisted removal of malachite green: Optimization, isotherm and kinetics. *RSC Advances, 5*(96), 79119–79128.

IEA Bioenergy. (2009). *Bioenergy—a sustainable and reliable energy source*. Paris, France: International Energy Agency Bioenergy.

Bi, X., Wang, P., Jiao, C., & Cao, H. (2009). Degradation of remazol golden yellow dye wastewater in microwave enhanced ClO_2 catalytic oxidation process. *Journal of Hazardous Materials, 168*(2–3), 895–900.

Bocos, E., Fernández-Costas, C., Pazos, M., & Sanromán, M.Á. (2015). Removal of PAHs and pesticides from polluted soils by enhanced electrokinetic-Fenton treatment. *Chemosphere, 125*, 168–174.

Borea, L., Naddeo, V., Shalaby, M. S., Zarra, T., Belgiorno, V., Abdalla, H., et al. (2018). Wastewater treatment by membrane ultrafiltration enhanced with ultrasound: Effect of membrane flux and ultrasonic frequency. *Ultrasonics, 83*, 42–47.

Bo, L., Zhang, Y., Quan, X., & Zhao, B. (2008). Microwave assisted catalytic oxidation of p-nitrophenol in aqueous solution using carbon-supported copper catalyst. *Journal of Hazardous Materials, 153*(3), 1201–1206.

Cecilia, R., Kunz, U., & Turek, T. (2007). Possibilities of process intensification using microwaves applied to catalytic microreactors. *Chemical Engineering and Processing: Process Intensification, 46*(9), 870–881.

Chen, W. X., Lee, J. Y., & Liu, Z. (2002). Microwave-assisted synthesis of carbon supported Pt nanoparticles for fuel cell applications. *Chemical Communications*, (21), 2588–2589.

Chen, J., Zhong, D., Hou, H., Li, C., Yang, J., Zhou, H., et al. (2016). Ferrite as an effective catalyst for HCB removal in soil: Characterization and catalytic performance. *Chemical Engineering Journal, 294*, 246–253.

Chipurici, P., Vlaicu, A., Calinescu, I., Vinatoru, M., Vasilescu, M., Ignat, N. D., et al. (2019). Ultrasonic, hydrodynamic and microwave biodiesel synthesis—a comparative study for continuous process. *Ultrasonics Sonochemistry, 57*, 38—47.

Dallinger, D., & Kappe, C. O. (2007). Microwave-assisted synthesis in water as solvent. *Chemical Reviews, 107*(6), 2563—2591.

Dastkhoon, M., Ghaedi, M., Asfaram, A., Goudarzi, A., Langroodi, S. M., Tyagi, I., et al. (2015). Ultrasound assisted adsorption of malachite green dye onto ZnS:Cu-NP-AC: Equilibrium isotherms and kinetic studies — response surface optimization. *Separation and Purification Technology, 156*, 780—788. https://doi.org/10.1016/j.seppur.2015.11.001

Dawei, L., Zhang, Y., Xie, Q., & Yazhi, Z. (2009). Microwave thermal remediation of crude oil contaminated soil enhanced by carbon fiber. *Journal of Environmental Sciences, 21*(9), 1290—1295.

Dhar, B. R., Nakhla, G., & Ray, M. B. (2012). Techno-economic evaluation of ultrasound and thermal pretreatments for enhanced anaerobic digestion of municipal waste activated sludge. *Waste Management, 32*(3), 542—549.

Ealias, A. M., & Saravanakumar, M. (2018). Facile synthesis and characterisation of AlNs using protein rich solution extracted from sewage sludge and its application for ultrasonic assisted dye adsorption: Isotherms, kinetics, mechanism and RSM design. *Journal of Environmental Management, 206*, 215—227.

Falciglia, P. P., Roccaro, P., Bonanno, L., De Guidi, G., Vagliasindi, F. G., & Romano, S. (2018). A review on the microwave heating as a sustainable technique for environmental remediation/detoxification applications. *Renewable and Sustainable Energy Reviews, 95*, 147—170.

Feng, Q. L., Wang, X. Q., Jia, Y., & Ning, P. (2013). Study on microwave combined with active carbon for the COD remove of amoxicillin wastewater. In *Paper presented at the applied mechanics and materials*.

Fetyan, N. A., & Salem Attia, T. M. (2020). Water purification using ultrasound waves: Application and challenges. *Arab Journal of Basic and Applied Sciences, 27*(1), 194—207.

Filly, A., Fernandez, X., Minuti, M., Visinoni, F., Cravotto, G., & Chemat, F. (2014). Solvent-free microwave extraction of essential oil from aromatic herbs: From laboratory to pilot and industrial scale. *Food Chemistry, 150*, 193—198.

Fitzgerald, M. E., Griffing, V., & Sullivan, J. (1956). Chemical effects of ultrasonics—"hot spot" chemistry. *The Journal of Chemical Physics, 25*(5), 926—933.

Flórez, N., Conde, E., & Domínguez, H. (2015). Microwave assisted water extraction of plant compounds. *Journal of Chemical Technology and Biotechnology, 90*(4), 590—607. https://doi.org/10.1002/jctb.4519

Ganzler, K., & Salgo, A. (1987). Microwave-extraction-a new method superseding traditional Soxhlet extraction. *Zeitschrift für Lebensmittel-Untersuchung und -Forschung, 184*(4), 274—276.

Ghaedi, M., Azad, F. N., Dashtian, K., Hajati, S., Goudarzi, A., & Soylak, M. (2016). Central composite design and genetic algorithm applied for the optimization of ultrasonic-assisted removal of malachite green by ZnO nanorod-loaded activated carbon. *Spectrochimica Acta Part A: Molecular and Biomolecular Spectroscopy, 167*, 157—164.

Ghaedi, M., Hajati, S., Zaree, M., Shajaripour, Y., Asfaram, A., & Purkait, M. (2015). Removal of methyl orange by multiwall carbon nanotube accelerated by ultrasound devise: Optimized experimental design. *Advanced Powder Technology, 26*(4), 1087—1093.

Gogate, P. R. (2017). Intensification of chemical processing applications using ultrasonic and microwave irradiations. *Current Opinion in Chemical Engineering, 17*, 9—14.

Gude, V. G. (2015). Synergism of microwaves and ultrasound for advanced biorefineries. *Resource-Efficient Technologies, 1*(2), 116—125.

Gude, V. G., & Grant, G. E. (2013). Biodiesel from waste cooking oils via direct sonication. *Applied Energy, 109*, 135—144. https://doi.org/10.1016/j.apenergy.2013.04.002

Gude, V. G., & Martinez-Guerra, E. (2015). Green chemistry of microwave-enhanced biodiesel production. In *Production of biofuels and chemicals with microwave* (pp. 225–250). Springer.

Gude, V. G., Patil, P., Martinez-Guerra, E., Deng, S., & Nirmalakhandan, N. (2013). Microwave energy potential for biodiesel production. *Sustainable Chemical Processes, 1*(1), 5. https://doi.org/10.1186/2043-7129-1-5

Guo, Z., Zhang, M., Li, H., Wang, J., & Kougoulos, E. (2005). Effect of ultrasound on antisolvent crystallization process. *Journal of Crystal Growth, 273*(3–4), 555–563.

Holkar, C. R., Jadhav, A. J., Pinjari, D. V., & Pandit, A. B. (2019). Cavitationally driven transformations: A technique of process intensification. *Industrial and Engineering Chemistry Research, 58*(15), 5797–5819.

Hong, Y.-D., Lin, B.-Q., Zhu, C.-J., & Li, H. (2016). Effect of microwave irradiation on petrophysical characterization of coals. *Applied Thermal Engineering, 102*, 1109–1125. https://doi.org/10.1016/j.applthermaleng.2016.04.019

Huang, K., Wang, S.-J., Shan, L., Zhu, Q., & Qian, J. (2007). Seeking synergistic effect—a key principle in process intensification. *Separation and Purification Technology, 57*(1), 111–120. https://doi.org/10.1016/j.seppur.2007.03.022

Hu, B., Wang, S.-B., Wang, K., Zhang, M., & Yu, S.-H. (2008). Microwave-assisted rapid facile "green" synthesis of uniform silver nanoparticles: Self-assembly into multilayered films and their optical properties. *Journal of Physical Chemistry C, 112*(30), 11169–11174.

Jamshidi, M., Ghaedi, M., Dashtian, K., Hajati, S., & Bazrafshan, A. A. (2016). Sonochemical assisted hydrothermal synthesis of ZnO: Cr nanoparticles loaded activated carbon for simultaneous ultrasound-assisted adsorption of ternary toxic organic dye: Derivative spectrophotometric, optimization, kinetic and isotherm study. *Ultrasonics Sonochemistry, 32*, 119–131. https://doi.org/10.1016/j.ultsonch.2016.03.004

Khaligh, A., Mousavi, H. Z., Rashidi, A., & Shirkhanloo, H. (2018). Nitrogen-modified nanoporous activated carbon from eucalyptus leaves for ultrasound-assisted removal of basic dyes using derivative spectrophotometric method. *Journal of the Serbian Chemical Society, 83*(5), 651–668.

Khanal, S. K., Grewell, D., Sung, S., & Van Leeuwen, J. (2007). Ultrasound applications in wastewater sludge pretreatment: A review. *Critical Reviews in Environmental Science and Technology, 37*(4), 277–313.

Kim, G.-D., Yoon, C., Kye, S.-B., Lee, Y., Kang, J., Yoo, Y., et al. (2012). A single FPGA-based portable ultrasound imaging system for point-of-care applications. *IEEE Transactions on Ultrasonics, Ferroelectrics, and Frequency Control, 59*(7), 1386–1394.

Knorr, D., Zenker, M., Heinz, V., & Lee, D.-U. (2004). Applications and potential of ultrasonics in food processing. *Trends in Food Science and Technology, 15*(5), 261–266.

Kostas, E. T., Beneroso, D., & Robinson, J. P. (2017). The application of microwave heating in bioenergy: A review on the microwave pre-treatment and upgrading technologies for biomass. *Renewable and Sustainable Energy Reviews, 77*, 12–27. https://doi.org/10.1016/j.rser.2017.03.135

Kumar, A., Gogate, P. R., Pandit, A. B., Wilhelm, A. M., & Delmas, H. (2005). Investigation of induction of air due to ultrasound source in the sonochemical reactors. *Ultrasonics Sonochemistry, 12*(6), 453–460.

Kumar, H., Lester, E., Kingman, S., Bourne, R., Avila, C., Jones, A., et al. (2011). Inducing fractures and increasing cleat apertures in a bituminous coal under isotropic stress via application of microwave energy. *International Journal of Coal Geology, 88*(1), 75–82.

Laugier, F., Andriantsiferana, C., Wilhelm, A.-M., & Delmas, H. (2008). Ultrasound in gas–liquid systems: Effects on solubility and mass transfer. *Ultrasonics Sonochemistry, 15*(6), 965–972.

Li, H., Qu, Y., Yang, Y., Chang, S., & Xu, J. (2016). Microwave irradiation–A green and efficient way to pretreat biomass. *Bioresource Technology, 199*, 34–41.

Li, H., Wu, C., Hao, Z., Li, X., & Gao, X. (2019). Process intensification in vapor—liquid mass transfer: The state-of-the-art. *Chinese Journal of Chemical Engineering, 27*(6), 1236—1246.

Luo, J., Fang, Z., & Smith, R. L. (2014). Ultrasound-enhanced conversion of biomass to biofuels. *Progress in Energy and Combustion Science, 41*, 56—93. https://doi.org/10.1016/j.pecs.2013.11.001

Lu, R., Yang, J., Xu, X., & Gao, J. (2009). Microwave-chemical desulphurization of sulfurous crude oil. *Petroleum Science and Technology, 27*(16), 1789—1799.

Lv, S., Chen, X., Ye, Y., Yin, S., Cheng, J., & Xia, M. (2009). Rice hull/$MnFe_2O_4$ composite: Preparation, characterization and its rapid microwave-assisted COD removal for organic wastewater. *Journal of Hazardous Materials, 171*(1—3), 634—639.

Ma, Y.-A., Cheng, Y.-M., Huang, J.-W., Jen, J.-F., Huang, Y.-S., & Yu, C.-C. (2014). Effects of ultrasonic and microwave pretreatments on lipid extraction of microalgae. *Bioprocess and Biosystems Engineering, 37*(8), 1543—1549.

Makino, K., Mossoba, M. M., & Riesz, P. (1983). Chemical effects of ultrasound on aqueous solutions. Formation of hydroxyl radicals and hydrogen atoms. *Journal of Physical Chemistry, 87*(8), 1369—1377.

Martinez-Guerra, E., & Gude, V. G. (2014a). *Fuel Synergistic effect of simultaneous microwave and ultrasound irradiations on transesterification of waste vegetable oil.*

Martinez-Guerra, E., & Gude, V. G. (2014b). Synergistic effect of simultaneous microwave and ultrasound irradiations on transesterification of waste vegetable oil. *Fuel, 137*, 100—108.

Martin, A., & Navarrete, A. (2018). Microwave-assisted process intensification techniques. *Current Opinion in Green and Sustainable Chemistry, 11*, 70—75.

Mary Ealias, A., & Saravanakumar, M. (2019). A critical review on ultrasonic-assisted dye adsorption: Mass transfer, half-life and half-capacity concentration approach with future industrial perspectives. *Critical Reviews in Environmental Science and Technology, 49*(21), 1959—2015.

Mason, T. J. (1999). Ultrasonically assisted extraction of bioactive principles from plants and their constituents. *Advances in Sonochemistry, 5*, 209—248.

Mason, T. J. (2012). Trends in sonochemistry and ultrasonic processing. In *Paper presented at the AIP conference proceedings.*

Mason, A., Korostynska, O., Ortoneda-Pedrola, M., Shaw, A., & Al-Shamma'a, A. (2013). A resonant co-planar sensor at microwave frequencies for biomedical applications. *Sensors and Actuators A: Physical, 202*, 170—175.

Ma, Y., Ye, X., Hao, Y., Xu, G., Xu, G., & Liu, D. (2008). Ultrasound-assisted extraction of hesperidin from Penggan (*Citrus reticulata*) peel. *Ultrasonics Sonochemistry, 15*(3), 227—232.

Midathana, V. R., & Moholkar, V. S. (2009). Mechanistic studies in ultrasound-assisted adsorption for removal of aromatic pollutants. *Industrial and Engineering Chemistry Research, 48*(15), 7368—7377. https://doi.org/10.1021/ie900049e

Motevali, A., Minaei, S., Banakar, A., Ghobadian, B., & Khoshtaghaza, M. H. (2014). Comparison of energy parameters in various dryers. *Energy Conversion and Management, 87*, 711—725.

Musielak, G., Mierzwa, D., & Kroehnke, J. (2016). Food drying enhancement by ultrasound—a review. *Trends in Food Science and Technology, 56*, 126—141.

Nadagouda, M. N., & Varma, R. S. (2008). Green synthesis of silver and palladium nanoparticles at room temperature using coffee and tea extract. *Green Chemistry, 10*(8), 859—862.

Neis, U. (2015). The use of power ultrasound for wastewater and biomass treatment. *Power Ultrasonics*, 973—996. Elsevier.

Nüchter, M., Ondruschka, B., Bonrath, W., & Gum, A. (2004). Microwave assisted synthesis—a critical technology overview. *Green Chemistry, 6*(3), 128—141.

Pakhomov, A. G., Prol, H. K., Mathur, S. P., Akyel, Y., & Campbell, C. B. G. (1997). Search for frequency-specific effects of millimeter-wave radiation on isolated nerve function. *Bioelectromagnetics, 18*(4), 324–334. https://doi.org/10.1002/(sici)1521-186x(1997)18:4<324::Aid-bem5>3.0.Co;2-4

Pal, A., Shah, S., & Devi, S. (2009). Microwave-assisted synthesis of silver nanoparticles using ethanol as a reducing agent. *Materials Chemistry and Physics, 114*(2–3), 530–532.

Paniwnyk, L. (2017). Applications of ultrasound in processing of liquid foods: A review. *Ultrasonics Sonochemistry, 38*, 794–806.

Park, B. K., Jeong, S., Kim, D., Moon, J., Lim, S., & Kim, J. S. (2007). Synthesis and size control of monodisperse copper nanoparticles by polyol method. *Journal of Colloid and Interface Science, 311*(2), 417–424.

Park, S.-J., & Yeo, S.-D. (2008). Recrystallization of caffeine using gas antisolvent process. *The Journal of Supercritical Fluids, 47*(1), 85–92.

Pilli, S., Bhunia, P., Yan, S., LeBlanc, R. J., Tyagi, R. D., & Surampalli, R. Y. (2011). Ultrasonic pretreatment of sludge: A review. *Ultrasonics Sonochemistry, 18*(1), 1–18. https://doi.org/10.1016/j.ultsonch.2010.02.014

Rai, S., Singh, U. P., Mishra, G. D., Singh, S. P., & Samarketu. (2009). Effect of water's microwave power density memory on fungal spore germination. *Electro-and Magnetobiology, 13*(3), 247–252. https://doi.org/10.3109/15368379409030722

Remya, N., & Lin, J.-G. (2015). Microwave-granular activated carbon (MW-GAC) system for carbofuran degradation: Analysis of characteristics and recyclability of the spent GAC. *Desalination and Water Treatment, 53*(6), 1621–1631.

Roosta, M., Ghaedi, M., & Yousefi, F. (2015). Optimization of the combined ultrasonic assisted/adsorption method for the removal of malachite green by zinc sulfide nanoparticles loaded on activated carbon: Experimental design. *RSC Advances, 5*(121), 100129–100141.

Saad, M., & Tahir, H. (2017). Synthesis of carbon loaded γ-Fe_2O_3 nanocomposite and their applicability for the selective removal of binary mixture of dyes by ultrasonic adsorption based on response surface methodology. *Ultrasonics Sonochemistry, 36*, 393–408.

Şahin, S., Samli, R., Tan, A. S. B., Barba, F. J., Chemat, F., Cravotto, G., et al. (2017). Solvent-free microwave-assisted extraction of polyphenols from olive tree leaves: Antioxidant and antimicrobial properties. *Molecules, 22*(7), 1056.

Shang, H., Zhang, H., Du, W., & Liu, Z. (2013). Development of microwave assisted oxidative desulfurization of petroleum oils: A review. *Journal of Industrial and Engineering Chemistry, 19*(5), 1426–1432.

Shokrollahi, F., Lau, K., & Tay, W. (2020). Performance comparison of ultrasonic-assisted and magnetic stirred absorption methods for CO_2 separation. *SN Applied Sciences, 2*(7), 1–10.

Singh, S., Gupta, D., Jain, V., & Sharma, A. K. (2014). Microwave processing of materials and applications in manufacturing industries: A review. *Materials and Manufacturing Processes, 30*(1), 1–29. https://doi.org/10.1080/10426914.2014.952028

Stankiewicz, A. (2006). Energy matters: Alternative sources and forms of energy for intensification of chemical and biochemical processes. *Chemical Engineering Research and Design, 84*(7), 511–521. https://doi.org/10.1205/cherd.05214

Sun, Y., Zhang, Y., & Quan, X. (2008). Treatment of petroleum refinery wastewater by microwave-assisted catalytic wet air oxidation under low temperature and low pressure. *Separation and Purification Technology, 62*(3), 565–570.

Tay, W., Lau, K., & Shariff, A. (2017). High frequency ultrasonic-assisted chemical absorption of CO_2 using monoethanolamine (MEA). *Separation and Purification Technology, 183*, 136–144.

Tsuji, M., Hashimoto, M., Nishizawa, Y., Kubokawa, M., & Tsuji, T. (2005). Microwave-assisted synthesis of metallic nanostructures in solution. *Chemistry—A European Journal, 11*(2), 440—452.

Van Gerven, T., & Stankiewicz, A. (2009). Structure, energy, synergy, time the fundamentals of process intensification. *Industrial and Engineering Chemistry Research, 48*(5), 2465—2474.

Vázquez-López, M., Amabilis-Sosa, L. E., Moeller-Chávez, G. E., Roé-Sosa, A., Neumann, P., & Vidal, G. (2019). Evaluation of the ultrasound effect on treated municipal wastewater. *Environmental Technology, 40*(27), 3568—3577.

Vinatoru, M., Mason, T. J., & Calinescu, I. (2017). Ultrasonically assisted extraction (UAE) and microwave assisted extraction (MAE) of functional compounds from plant materials. *TRAC Trends in Analytical Chemistry, 97*, 159—178.

Wang, H., Ding, J., & Ren, N. (2016). Recent advances in microwave-assisted extraction of trace organic pollutants from food and environmental samples. *TRAC Trends in Analytical Chemistry, 75*, 197—208.

Wang, N., & Wang, P. (2016). Study and application status of microwave in organic wastewater treatment—a review. *Chemical Engineering Journal, 283*, 193—214.

Wang, Y., Yu, J., Xiao, W., & Li, Q. (2014). Microwave-assisted hydrothermal synthesis of graphene based Au—TiO_2 photocatalysts for efficient visible-light hydrogen production. *Journal of Materials Chemistry, 2*(11), 3847—3855.

Wu, L., Hu, M., Li, Z., Song, Y., Yu, C., Zhang, H., et al. (2016). Dynamic microwave-assisted extraction combined with continuous-flow microextraction for determination of pesticides in vegetables. *Food Chemistry, 192*, 596—602.

Xu, D., Cheng, F., Zhang, Y., & Song, Z. (2014). Degradation of methyl orange in aqueous solution by microwave irradiation in the presence of granular-activated carbon. *Water, Air, & Soil Pollution, 225*(6), 1—7.

Yuan, D., Fu, D. Y., & Luo, Z. W. (2013). Study on microwave combined with granular active carbon for treatment of p-nitrophenol wastewater. In *Paper presented at the advanced materials research.*

Yusof, S. M. M., Shariff, A. M., Tay, W. H., Lau, K. K., & Mustafa, N. F. A. (2020). Mass transfer intensification of CO_2 absorption in monoethanolamine using high frequency ultrasonic technology in continuous system. *International Journal of Greenhouse Gas Control, 102*, 103157.

Zhang, H., Duan, L., & Zhang, D. (2007). Absorption kinetics of ozone in water with ultrasonic radiation. *Ultrasonics Sonochemistry, 14*(5), 552—556.

CHAPTER TWO

Sorption processes under high-gravity field

2.1 Introduction

Process intensification (PI) involves novel processing techniques and/or equipment, often based on alternative forms of energy input aimed at enhancing process performance and at reducing capital and/or operating costs (Becht, Franke, Geißelmann, & Hahn, 2009; Nikačević, Huesman, Van den Hof, & Stankiewicz, 2012; Stefanidis & Stankiewicz, 2013). Imperial Chemical Industries (ICI) have introduced the concept of PI in the 1970s to define the reduction in capital cost of an existing production system deriving from physical miniaturization of process equipment and/or application of innovative separation techniques (Ponce-Ortega, Al-Thubaiti, & El-Halwagi, 2012). HiGee is one of the practical ways of applying PI. This device, which relies on centrifugal force to replace gravity in facilitating fluid flow through a packed bed, thus increasing the effectiveness of transport phenomena in the bed, was invented in 1979 by Colin Ramshaw (Lin & Chu, 2015). Due to its characteristics, HiGee can operate at higher gas and/or liquid flow rates owing to the lower tendency to flooding and with mass transfer rate enhancement by a factor of 10−100, leading to a dramatic reduction in equipment size, and therefore in both capital and operating costs (Chen, Lin, & Liu, 2005; Lin, Wei, Hsu, & Liu, 2006). The utilization of HiGee has been well known in many applications, such as absorption, distillation, adsorption, synthesis, mixing and segregation, precipitation, and stripping (Li et al., 2013; Lin & Chien, 2008). Therefore, due to its important role in process intensification, this chapter has focused on a review of this technology starting from its fundamental principles and including its applications for different processes. The results of this analysis can be beneficial for anyone who are interested in research in and/or application of process intensification field and its dependent area.

One of the chemical engineering fields which HiGee can be used for intensification of the process is sorption processes. Sorption process is one of the most commonly used separation technologies; by using which one

Intensification of Sorption Processes
ISBN: 978-0-12-821411-4
https://doi.org/10.1016/B978-0-12-821411-4.00007-9

substance becomes attached to another. It should be noted that the reverse of sorption is desorption. Sorption can be completely physical or accompanying with reaction, leading to chemical sorption. Specific cases of sorption are classified as follows:

- Absorption:
  Absorption is the incorporation of a substance in one state into another of a different state. For example, the liquids being absorbed by a solid or the absorption of gases by a liquid.
- Adsorption:
  Adsorption is the physical bonding or adherence of molecules or even ions onto the surface of another phase. For example, the adsorption of reagents onto a solid catalyst surface.
- Ion exchange:
  Ion exchange is another types of sorption where an exchange of ions between two electrolytes or between an electrolyte solution and a complex is performed.

Innovative separation concepts believed as a process intensification offering reduction in costs and the size of equipment, improvement in product quality, and enhancing the yields of sorption process, have remarkably received attention in recent years. One of the alternative technologies in separation processes especially the sorption ones is the HiGee technology, the essence of which is replacing the gravitational field with a high centrifugal field. HiGee technology improves the performance of mass transfer processes by providing high-gravity field, increases separation efficiency and capacity at a very compact design. HiGee contactors operate in a high centrifugal field up to even 1000 times more than gravitational field (Fig. 2.1). It brings about up to enhancement of one to two orders of magnitude in mass transfer leading to drastic reduction in size with respect to conventional equipment.

RPBs can be classified based on the different factors, such as:

- contacting the phases (cocurrent, countercurrent, and crosscurrent)
- stages (single stage or multiple stages)
- number of rotor(s) (single rotor or multiple rotors)
- rotor type (single block or split packing)
- rotor orientation (vertical or horizontal)
- packing type (conventional packings, blade, blade and baffle, concentric baffle, concentric rings, zig-zag, ….)
- electrical motor to rotor shaft coupling (direct and indirect)

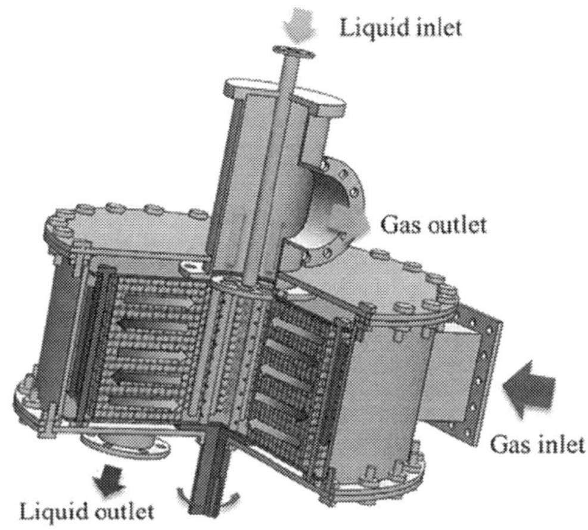

Figure 2.1 Schematic diagram of a HiGee contactor for gas absorption (Xie et al., 2019).

HiGee technology has found numerous application in three areas:

- Engineering of nanoparticles
- Reaction process intensification
- Separation process intensification

In this chapter, the HiGee technology focusing on the sorption process is discussed. Different aspects including the sorption mechanism, the HiGee equipment, the economic evaluation, the advantages and drawbacks will be elaborated.

2.2 Fundamental principles and operating characteristics

In this section, the performance of the rotating packed bed (RPB) is evaluated as an efficient HiGee equipment for intensification of sorption process. RPB essentially applies a high-gravity environment, caused by the significant centrifugal force (100−1000*g*) generated by a high rotational speed, to mass transfer in a packed bed. The details of a typical RPB are shown in Fig. 2.2.

The increased gravity creates greater shear force, increases turbulence and micromixing, and leads to the splitting of relatively thick liquid films into tiny droplets and thin-film layers (Chiu, Chen, & Huang, 2007; Lin & Chen, 2011a; Tan & Lee, 2008; Zhang et al., 2012). The packing materials usually used in RPB applications also have large surface area (1000−4000 m^2/m^3), which provides 5−10 times higher mass transfer rates compared to conventional packed beds (Jiao, Liu, & Qi, 2010a; Shivhare, Rao, & Kaistha, 2013). Therefore, due to both centrifugal force and higher

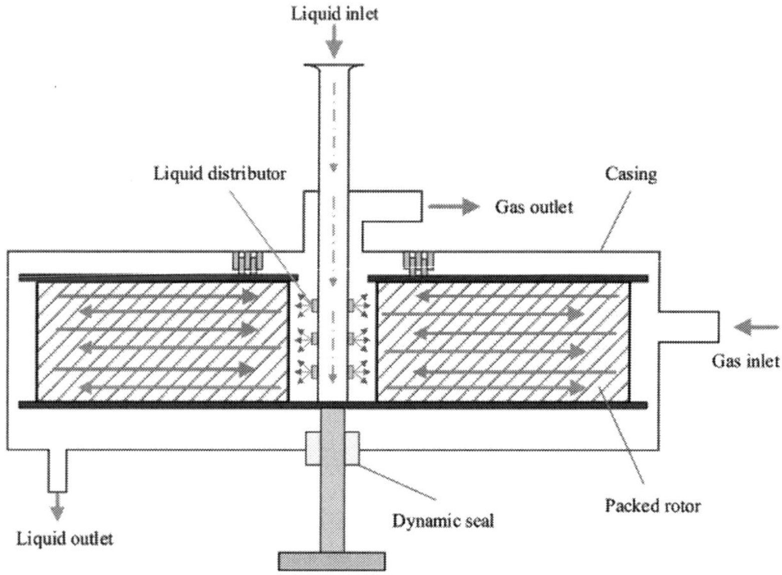

Figure 2.2 Sketch of a typical RPB (Zhang, Xie, Li, Teng, & Zhu, 2020).

interfacial surface area, mass transfer is dramatically enhanced. For this reason RPBs have very low height transfer units (1−2 cm), and micromixing (i.e., mixing at the molecular scale) can be up to one to three orders of magnitude larger than that in conventional packed beds (Zhao, Shao, and Chen, 2010). Additionally, RPBs can be operated at much higher gas or liquid flow rates, due to lower tendency to flooding compared to conventional packed bed. This aspect too leads to a reduction in physical size, and as a net result process capital and operating costs are significantly reduced (Lin, Chao, & Liu, 2010; Lin, Chao, Liu, & Lee, 2009). The investigation of RPB operating characteristics and understanding its hydrodynamic is very necessary and has an important role in scale-up for commercial applications which has pointed in next section.

2.2.1 Pressure drop

One of the most important operational parameter that directly affects possible flow rates as well as cost of process is the pressure drop. The total pressure drop in RPB is attributed to the sum of the pressure drop inside the rotor and the pressure drop outside the rotor (Keyvani & Gardner, 1989). Rotational speed and gas and liquid flow rates are the main factors that influence on the pressure drop in RPB (Lin & Chen, 2008; Xiuping & Youzhi, 2010). According to literature results, in both "dry" and "wet" RPBs, the pressure drop increases with the square of the rotor speed at constant gas and liquid flow rates (Lin, Chen, & Liu, 2000; Wang et al., 2008). On the other hand, an increase in gas or liquid flow rates causes an increase in pressure drop, with a smaller effect attributable to liquid flow rate (Sung & Chen, 2012). Jiao et al. testing different packing structures, proposed the following correlations for "dry" and "wet" pressure drop (Jiao et al., 2010a):

$$\Delta P_d = A' \beta^a G^b \tag{2.1}$$

$$\Delta P_w = A \beta^a G^b L^c \tag{2.2}$$

where ΔP_d and ΔP_w are pressure drop in dry and wet bed, G and L are gas and liquid flow rate and A, A', a, b, and c are coefficients. Based on their report, good agreement between experimental data and predicted results was observed. Correlations for the pressure drop are listed in Table 2.1, which indicate that pressure drop in RPB can be considered as a function of some operational parameters as well as, mechanical characteristics such as: fluid velocity, gas and liquid properties, centrifugal force, packing characteristics, and fractional resistance.

2.2.2 Liquid holdup

The liquid holdup behavior in RPBs and conventional equipment is significantly different. In general, liquid and gas holdup is lower than that in conventional packed bed, leading to increased safety and, for the same reason, reduced costs as a results of smaller size of equipment. Since the smaller devices are fabricated using expensive and durable materials, they can be safer (Sung & Chen, 2012). Measurement of liquid holdup in rotating packed bed is very difficult, so many investigation focused on this problem, aiming at individuating a satisfactory technique (Lin et al., 2000). Lin et al. (2000) reported that liquid holdup, h_L, in RPB can be predicted with good accuracy using the following equation, in which a modified Galileo number is used.

$$h_L = 3.86 (Re_{Ld})^{0.545} \left(Ga^*\right)^{-0.42} \left(\frac{a_t d_p}{\varepsilon}\right)^{0.65} \varepsilon \tag{2.3}$$

where Re_{Ld} is liquid Reynolds number, d_p is spherical equivalent diameter of the packing, a_t is total specific surface area of the packing, and ϵ is voidage.

Table 2.1 Correlations of pressure drop prediction in RPB.

RPB type	Correlation	Reference
Countercurrent flow	$\Delta P = 0.09\rho_G\omega^2\left(R_2^2 - R_1^2\right) + \frac{1.7(R_2-R_1)G'^2}{\rho_G R_k} Re_G^{-0.58516} Re_L^{0.58763}$	Hao and Nangeng (1994)
Countercurrent flow	$\Delta P = \frac{1}{2}\rho_G(k_s\omega)^2\left(R_2^2 - R_1^2\right) + \frac{1}{2}\left(k_{af}\right)R_2^2 Q_G^2 + \frac{a_f(1-\varepsilon)}{\rho_G\varepsilon^2}\Delta I$	Kumar and Rao (1990)
Zigzag-flow	$\Delta P = \xi_i\frac{\rho_G u_i^2}{2} + \Delta p_{zi} + \sum_{j=1,3,5}^{2n-1}\Delta p_{j\to j+1} + \xi_o\frac{\rho_G u_o^2}{2}$	Sui, Xu, Yu, and Ji (2008)
Countercurrent flow	$\Delta P = \frac{1}{2}\rho_G K V_1^2 + \frac{1}{2}\rho_G\left(V_e^2 - V_1^2\right) + \frac{1}{2}\rho_G\left(\frac{Q_g}{2\pi a\varepsilon}\right)^2\left(\frac{1}{r_i^2} - \frac{1}{r_o^2}\right)$ $+\rho_G\int_{r_i}^{r_o}\frac{\overline{V}_\theta^2}{r}d + \frac{\rho_G}{2\varepsilon^2 d_h}\left(\frac{Q_g}{2\pi a}\right)^2\left[a\frac{2\pi a v_g}{Q_g d_h}\ln\frac{r_0}{r_i} + \varnothing\beta\left(\frac{1}{r_i} - \frac{1}{r_o}\right)\right]$	Sandilya, Rao, Sharma, and Biswas (2001)
Countercurrent flow	$\Delta P = \frac{F_V^2}{2}\frac{a_t}{\varepsilon - h_L} f_W\left(R_o - R_i\right)$	Lin and Jian (2007)

Since the smaller devices are fabricated using expensive and durable materials, they can be safer.

Determination of liquid hold-up using electrical resistance method carried out by Burns et al. (Burns, Jamil, & Ramshaw, 2000) and showed that hold-up was independent of gas flow up to the flooding point, and the dependence of hold-up to liquid viscosity is weak. Based on the result, the best fit hold-up correlations was obtained as follows:

$$h_L = 0.039\left(\frac{g}{g_0}\right)^{-0.5}\left(\frac{U}{U_0}\right)^{0.6}\left(\frac{\nu}{\nu_0}\right)^{0.22} \tag{2.4}$$

where g is centrifugal acceleration, g_0 is a characteristic centrifugal acceleration, U is the liquid flow rate per unit area (superficial velocity), U_0 is a characteristic flow rate per unit area, ν is kinematic viscosity of the liquid, and ν_0 is a characteristic kinematic viscosity.

2.2.3 Flooding

In conventional packed beds it is well known that when flow increases, liquid movement through the packed bed becomes more difficult so liquid accumulates within the column and interfacial area decreases: this phenomenon is called flooding, and causes an unacceptable reduction in mass transfer between phases. According to studies, flooding correlation for RPB is similar to conventional packed beds when the gravity term was replaced by centrifugal acceleration (Ehinmowo et al., 2015; PMNT, 2004). Fluid behavior can be explained by the Sherwood flooding chart (Fig. 2.3), which is the plot of $\frac{u_g^2 a}{g\varepsilon^3}\frac{\rho_g}{\rho_L}$ versus $\frac{L}{G}\left(\frac{\rho_g}{\rho_L}\right)^{\frac{1}{2}}$. As acceleration increases, the gas flooding velocity (u_g) increases in order to maintain the same value of the first term. Since the ratio of liquid (L) to gas (G) flow remains constant, liquid flow increases commensurately with gas flow (PMNT, 2004). The RPB could be operated at higher gas and liquid flow rates due to low tendency to flooding compared to that in conventional packed bed as a results of action of centrifugal acceleration (Joel, Wang, & Ramshaw, 2015; Rao, 2015).

2.2.4 Residence time distribution

The residence time distribution (RTD) for flow in rotating packed bed is defined as a probability distribution for the time of a given element of fluid to pass through that RPB (Reay et al., 2013). RTD can be used to investigate fluid motion and mixing in continuous flow systems, and is an essential

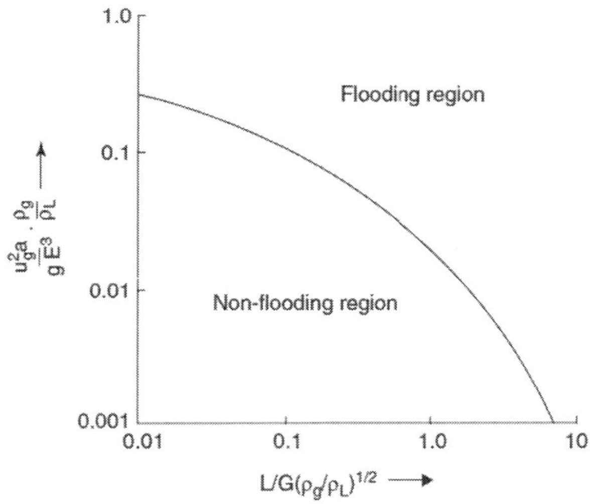

Figure 2.3 Sherwood flooding chart for gas–liquid flow in packed beds (Reay, Ramshaw, & Harvey, 2013).

operational factor for design and scaling up (Guo et al., 2000; Keyvani & Gardner, 1989). Based on the reports, the mean RTD on order of 0.4–2s were achieved for RPB as the rotational speeds were varied from 400 to 1200 rpm (Keyvani & Gardner, 1989). As the rotational speed or liquid flow rate increases, the mean RTD decreases (Lin et al., 2000). Literature results indicate that gas flow rate and liquid viscosity have little influence on RTD (Burns et al., 2000). RPBs have lower residence time compared to conventional equipment, hence less chemical inventory during processing (Hassan-Beck, 1997). RTD measurement in RPBs have been carried out using tracer methods, visual and electrical conductivity sensors (Burns et al., 2000; Lin et al., 2000; PMNT, 2004). Guo et al. studied fluid behavior to evaluation of film thickness, velocity profile and RTD using a high speed video camera as well as electroconductivity sensors mounted on the RPB (Guo et al., 2000). Analysis and modeling of results indicated that average residence time of liquid depended on liquid flow rate and rotating speed, and varies from 200 ms to 800 ms.

2.2.5 Micromixing and mass transfer

In RPBs the high gravity environment creates high shear and turbulence, leading to higher mixing at the molecular scale, known as micromixing. More intense micromixing is an advantage of RPBs, which can lead to improved selectivity, yield and quality of final products in industrial processes, such as polymerization, combustion, precipitation, crystallization and competing fast chemical reactions (Jiao, Liu, & Qi, 2010b; Yang, Chu, Yang, & Chen, 2006; Yang, Chu, Shao, Luo, & Chen, 2009). If the time scale of the chemical reaction involved is at the same magnitude as or smaller than the time scale of the mixing process, micromixing is a very significant operational parameter (Zhao et al., 2010). As a results of micromixing, mass transfer increases due to a created large contact area between phases, with an enhancement factor up to 10−100. Dues to this enhancement, RPB can be applied to many industrial processes such as absorption, adsorption, stripping, distillation, desorption, nanoparticles production and reactive crystallization (Lin & Chen, 2008). Thus, based on the above mentioned, development of experimental and theoretical methods in order to characterize micromixing and mass transfer in RPB is an essential issue. Mass transfer coefficients for rotating packed beds have been studied based on different theories such as penetration theory, surface renewal theory, and film theory (Luo et al., 2012; Munjal, Dudukovć, and Ramachandran, 1989a,b). Based on these theories, numerous different correlation equations for mass transfer coefficients for different processes including absorption, stripping and distillation have been developed (Boodhoo, Vicevic, Cartwright, Ndlovu, & Toogood, 2008; Guo, Zhang, Luo, Dang, & Qian, 2014; Guo, Zheng, Guo, Feng, & Gardner, 1997; Sung and Chen 2012; Yang et al. 2011). Most of correlation equations are expressed in terms of gas and liquid side volumetric mass transfer coefficients (k_Ga and k_La). Alternatively, height transfer unit (HTU), number of transfer units (NTU), area transfer unit (ATU), and volume transfer unit (VTU) are other techniques which have been often adopted for evaluating RPB performance (Xiuping & Youzhi, 2010; Lin & Jian, 2007; PMNT, 2004). The correlations relative to mass transfer based on volumetric mass transfer coefficients were detailed in Table 2.2.

2.3 Computational fluid dynamics (CFD) and mathematical modeling studies

The experimental measurements which are required to understand RPB behavior in different process require special devices and different RPBs structures, which is very time-consuming and costly (Shi, Yang, Wen, & Chen, 2013; Yang, Wang, Chen, & Fei, 2010). Furthermore, visual and experimental studies often cannot deliver comprehensive information about important characteristic details, such as flow patterns, phase

Table 2.2 Correlations for prediction of volumetric mass transfer coefficient in RPB.

Experimental system	Correlation	Reference
Ethanol-water	$\frac{k_G a R T}{D_G a_t^2} = 3.111 \times 10^{-3} Re_G^{1.163} Re_L^{0.631} Gr_G^{0.25}$	Liu, Lin, Wu, and Hsu (1996)
Ammonia-water	$k_G = \left(k_s D_G u_G^2\right)^{0.5}$	Guo et al. (1997)
VOCs-water	$\frac{k_G a H^{0.27} R T}{D_G a_t^2} = 0.077 Re_G^{0.323} Re_L^{0.328} Gr_G^{0.18}$	Chen and Liu (2002)
Isopropyl alcohol/ethyl acetate-Water	$\frac{k_G a H^{0.315} R T}{D_G a_t^2} = 0.061 Re_G^{0.712} Re_L^{0.507} Gr_G^{0.326}$	Lin, Wei, Liu, and Shen (2004)
Oxygen/isopropyl-Water	$\frac{k_G a d_p}{D_G a_t} = 0.027 Re_G^{0.679} Re_L^{0.245} Gr_G^{0.140}$	Lin and Jian (2007)
Ammonia/isopropyl alcohol-Water, isopropyl alcohol/acetone/ethyl acetate-water	$\frac{k_G a H^{0.185} R T}{D_G a_t^2} = 0.0186 Re_G^{0.389} Re_L^{0.534} Gr_G^{0.245}$	Chen, Hsu, Lin, Tai, and Liu (2008)

distribution and, more in general, fluid dynamics features. On the other hand, simulation and modeling based on CFD or mathematical methods are powerful tools for prediction and analyzing the hydrodynamic characteristics and other phenomena such as mass transfer in rotating packed bed system leads to improvement in designing of an RPB with higher performance. Shi et al. developed a two dimensional CFD model for analysis of liquid phase flow in an RPB reactor (Shi et al., 2013). The influence of rotational speed and structure of packings on the velocity, size distribution and residence time of the liquid phase were studied, and the results showed that increasing in rotational speed enhances liquid velocity in packing area. Furthermore, as rotational speed increases, droplet diameter and mean residence time decreases. A 3D unsteady-state single-phase flow simulations were carried out by Llerena-Chavez et al. for investigation of dry pressure drop and gas flow maldistribution in three different RPB designs. Their analysis allowed to individuate better configuration of gas inlet in view of pressure drop reduction and flow distribution improvement (Llerena-Chavez & Larachi, 2009). CFD simulation for better understanding of fluid behavior in RPB was performed for multiphase flow of water-SO_2 system. Prediction and description of velocity distribution in the domain of interest, pressure drop, and flow pattern using a three-dimensional model was also performed (Martínez, Jaimes, Gomez, & Filho, 2012).

Mathematical modeling and correlations based on empirical data and/or theories were performed to understanding of RPB. CO_2 absorption process by monoethanolamine solution modeling was carried out by Kang et al. using two-film theory for mass transfer (Kang, Sun, Wong, Jang, & Tan, 2014). The effects of gas and liquid flow rates, rotational speed, and solvent conditions on absorption efficiency were investigated and comparison of result with literature reports indicated that proposed model had good ability for prediction. Surface coverage model based on response surface methodology was applied to modeling and optimization of kinetics of carbonation reaction of basic oxygen furnace slags in RPB (Pan, Chiang, Chen, Tan, & Chang, 2014). Results reveals that the overall maximum carbonation conversion of basic oxygen furnace slags was found to be 95% under optimum conditions as rotational speed of 840 rpm and a slurry flow rate of 0.85 L/min at 65 C and time of 30 min. Furthermore, Modeling of CO_2 absorption in RPB for prediction of mass transfer performance and optimization of process was performed using Aspen Plus, artificial neural network, and other techniques (Joel et al., 2015; Yi, Zou, Chu, Shao, & Chen, 2009; Zhao, Su, and Tao, 2014).

2.4 Intensification of the sorption processes using HiGee equipment

2.4.1 Absorption and stripping

Absorption and stripping are among the common processes in many industries such as gas refineries, in particular many researchers studied the application of RPB to carbon dioxide (CO_2) absorption, in light of the high significance of this process for greenhouse gases emissions control (Rahimi & Mosleh, 2015; Yu, Cheng, & Tan, 2012). Lin et al. studied the performance of the crossflow RPB in removing CO_2 using different absorbent solutions to evaluate of influence of various parameters on removal efficiency (Lin & Chen, 2011b). According to the experimental results, efficiency increased with rotational speed and liquid flow rate, but decreased with gas flow rate. Cheng et al. utilized an RPB system for absorption of CO_2 using an aqueous solution containing 30 wt.% of either monoethanolamine (MEA) or diethylenetriamine (DETA) mixed with piperazine (PZ) (Cheng & Tan, 2011). Results indicated that initial CO_2 concentration could be reduced from 1000 ppm to a level much less than 100 ppm in this system. Analysis of results based on the overall mass transfer coefficient (K_Ga) and on HTU showed that RPB had higher efficiency compared with a conventional packed bed, with K_Ga higher than 5.8 s^{-1} of and HTU smaller than 1.0 cm. According to our previous studies and many other researches, CO_2 absorption performance using different design of RPB was phenomenal (Cheng &Tan, 2009; Kang, Luo, et al., 2014; Lin & Chen, 2011a; Rahimi, Karimi, & Mosleh, 2013; Rahimi & Zarei, 2016; Yu & Tan, 2014). Furthermore, absorption of other compounds such as sulfur dioxide (SO_2), NO_x, and ozone was studied using RPB (Jiang, Youzhi, & Meiduo, 2011; Lin et al., 2010; Yu, Liu, Zhang, Su, & Jin, 2010).

Absorption of volatile organic compounds (VOCs) is another application for which RPBs have been proposed (Rahimi & Mosleh, 2013). VOCs are widely used as solvents and detergents in chemical industry, and emission of these compounds is very harmful environmentally. Absorption of methanol and 1-butanol from binary mixtures of VOCs was investigated using RPB (Hsu & Lin, 2011). The estimation of overall volumetric gas phase mass transfer coefficients (K_Ga) was used to evaluate the effects of operational parameters on mass transfer performance. Results showed that mass transfer is essentially controlled by gas side resistance. Removal of acetone from a viscous liquid using RPB was carried out with an efficiency about 67%

higher than that of a conventional devolatilizer due to reduction in mass transfer resistance achieved by increase of the generated surface renewal rate (Woyuan et al., 2009). RPB also was applied to absorption of other VOCs such as ethanol, ethyl acetate, isopropyl alcohol, methanol, methyl ethyl ketone and methyl acetate (Lin & Chien, 2008; Lin et al., 2006; Lin, Lin, & Chien, 2009). The RPB system can also be used for striping. Lin et al. studied the performance of an RPB system for the stripping of oxygen from water (Lin & Liu, 2007). In their work, the overall volumetric mass transfer coefficient (K_La) was evaluated as a function of the rotor speed, gas flow rate, and liquid flow rate. Results indicated that K_La increased with rotor speed and liquid flow rate, but was independent of the gas flow rate. Furthermore, once again extremely low values of HTU (1.9 ~ 2.3 cm) were observed, with a definite, RPB had more efficient mass transfer compared to that of a conventional packed bed.

2.4.2 Solvent regeneration

The energy cost of solvent regeneration is an economic concern in some industries, and RPB can be used for reduction of operational and capital cost of such processes. Results of experimental studies for regeneration of 30 wt% MEA aqueous solution that was loaded with CO_2 indicated that RPB had higher regeneration efficiency and less energy consumption compared to conventional packed bed (Cheng, Lai, & Tan, 2013). Similarly, experiments for thermal regeneration of alkanolamine using RPB system revealed that RPB provided higher volumetric mass and heat transfer coefficients and therefore possesses a lower HTU, so RPB had smaller size and volume related to conventional packed bed leads to saving costs (Cheng et al., 2013).

2.4.3 Adsorption

RPB can intensify adsorption processes by reducing mass transfer resistance. Removal of chromium by activated carbon was investigated in RPB and effect of operation parameters such as rotational speed, flow rate, pH and packing density on process efficiency were studied (Kundu, SenGupta, Hashim, & Redzwan, 2015). Based on the results, 91% removal efficiency was obtained at optimum conditions. Biosorption of Cu (II) ion using a continuous RPB was carried out, the results indicated that increasing in rotational speed leads to a very significant enhancement in volumetric mass transfer coefficient (Das, Bhowal, & Datta, 2008). RPB was also used for absorption of methomyl, a pesticide, on activated carbon. Based on the experimental analysis, due to high centrifugal force caused by high rotational speed, shorter time to reach equilibrium, and faster internal mass transfer were achieved (Chang & Lee, 2012).

2.4.4 Photocatalytic degradation

Photocatalytic degradation, an advanced oxidation process (AOP), is a clean and green purification technology in the wastewater treatment process (Mosleh & Rahimi, 2016). Despite conventional methods wherein pollutants are separated or transferred to another phase, in AOP pollutants are directly converted to carbon dioxide and water without further separation by producing strong oxidants (hydroxyl, •OH) (Mosleh, Rahimi, Ghaedi, & Dashtian, 2016). In our previous work, a novel RPB (Fig. 2.4) was used for intensification of photocatalytic degradation of pollutants with higher efficiency in comparison to conventional methods (Mosleh, Rahimi, Ghaedi, Dashtian, & Hajati, 2016b; Mosleh et al., 2017). Due to higher interfacial surface area created by high gravity environment, solid–liquid mass transfer between the photocatalyst particles and pollutant solution dramatically increased, and the same happened to photodegradation efficiency. The performance of rotating packed bed reactor was compared with a conventional one at same conditions, and it was shown that the

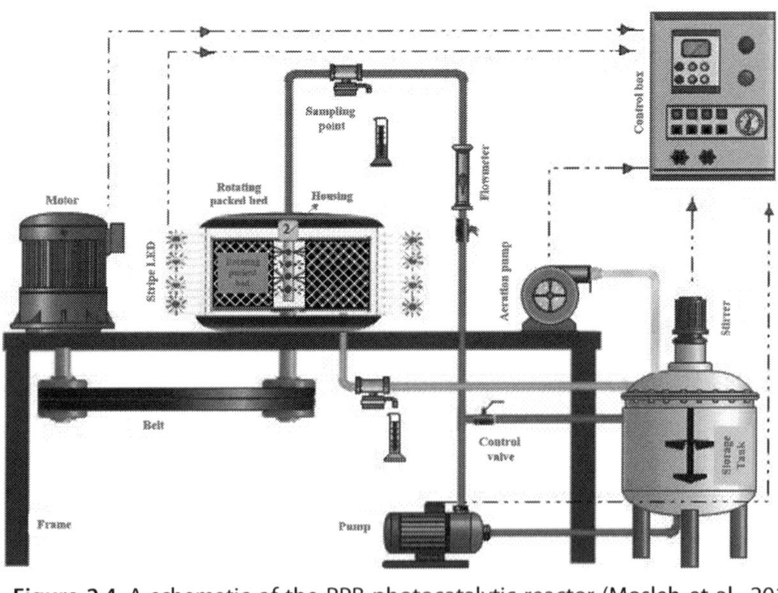

Figure 2.4 A schematic of the RPB photocatalytic reactor (Mosleh et al., 2016a).

RPB reactor is more efficient because of its lower consumption of photocatalyst and shorter illumination time (Mosleh, Rahimi, Ghaedi, Dashtian, & Hajati, 2016a).

2.4.5 Major industrial applications

Table 2.3 lists a major industrial application of high-gravity field separation processes in chemical plants.

Despite the industrial applications listed in Table 2.3, the deployment of the HiGee technology is still at the beginning of its way. The main reasons seem to be the concerns about the reliabilities and energy consumption of the rotating parts used in the separation equipment in this area. However, this concern should also be for lots of other types of the equipment like pumps and fans that have rotating elements in their structure, while all of which are widely used in the industry (Ramshaw, 1993).

Table 2.3 A list of industrial application of high-gravity field separation processes (Cortes Garcia, van der Schaaf, & Kiss, 2017).

Company	Application	Capacity	Year	Remarks
China Petrochemical Corporation/ Shengli oil field co. China	Seawater deaeration	Two units of 250 L/h	1998	• investment cost 40% lower • ground space reduced by 60% • equipment weight reduced by 80% • improved oxygen removal efficiency
The Dow chemical company, USA	Reactive stripping of HOCl	150 L/h	1999	• yield 10% higher • less than half stripping gas • investment cost 70% lower • operating cost 30% lower • reactor volume 40% smaller
Zibo Sulfuric acid plant (Shandong)	SO_2 absorption from	3000 m^3/h of gas	1999	• degree of equilibrium absorption close to 100%

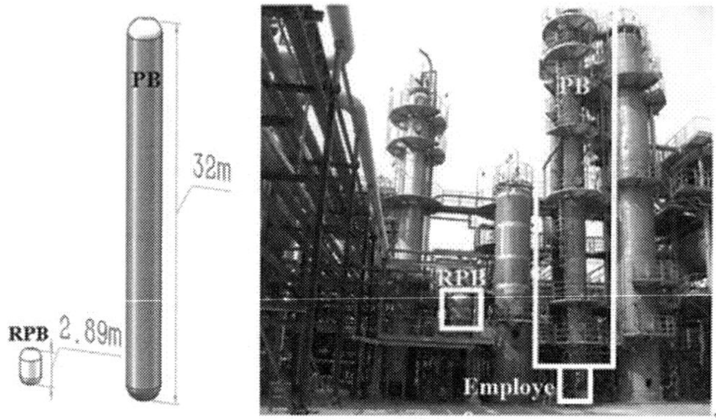

Figure 2.5 The difference in the size of the RBP and a conventional packed bed (Qian, Chen, & Grossmann, 2018).

RBP offers significant improvements in terms of energy efficiency and sustainability. The high flexibility, intense mass transfer and very compact design are among the advantages of application of RBP resulting in potential savings in operation and investment costs. Using the RBP dramatically reduced the size of the packed bed. In a study by Lin and Chen (2011a), the size of conventional packed bed and crossflow RBP were compared for the adsorption of CO_2. They demonstrated that for the same adsorption efficiency, the height of the conventional packed bed is 10 times larger than the crossflow RBP. Fig. 2.5 demonstrates the difference between the size of RBP and a conventional packed bed in a same place in a certain plant.

2.5 Other HiGee-based devices

2.5.1 Thin-film spinning disc reactor

The thin-film spinning disc reactor was commercially developed in 1997 (Jachuck et al., 1997). The design of this reactor is in such a way that at the center of the rotating disc the liquid phase is fed. After entering the reactor, the liquid phase flows radially outwards over the disc as a thin film. The thickness and radius of the liquid film vary with changing the rotational disk speed. In addition, three different parameters can increase the radial velocity:

- Increasing the rotational disk speed
- Increasing the liquid flow rate
- Decreasing the viscosity of the liquid phase

This type of the reactor has found several applications including crystallization, polymerization and the production of active pharmaceutical ingredient. Moreover, the thin-film spinning disc reactors can be especially exploited in the processes with very fast reactions and need large heat dissipation like sulfonations or exothermic condensation (Boodhoo & Al-Hengari, 2012).

2.5.2 Rotor-stator spinning disc reactor

The rotor-stator spinning disc reactor is the modified version of thin-film spinning disc reactor (Stakić & Urošević, 2011). As shown in Fig. 2.6, the structure of the rotor-stator spinning disc reactor differs from the thin-film spinning disc reactor. This reactor consists of a rotor, which is located between two stators (discs). The gas phase is fed through the bottom stator. As the gas phase enters, the gas bubbles are sheared off as a result of the presence of the shear stress in the cavity between the rotor and the stator. Note that the diameter of the gas bubble is the function of the rotational disc speed.

The gas-liquid mass transfer rate of the rotor-stator spinning disc reactors are influenced by the following parameters (Meeuwse, van der Schaaf, & Schouten, 2010):

- Rotational speed
- Rotor diameter
- Rotor-stator distance (The axial distance between the rotor and stators)
- The feed location of the gas phase

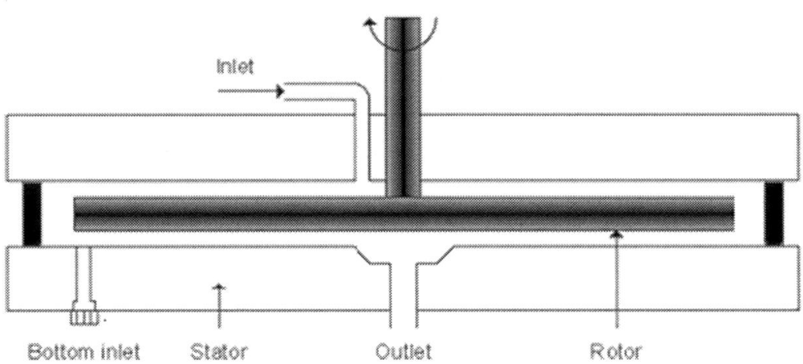

Figure 2.6 Side view of the rotor-stator spinning disc reactor (Visscher, Van der Schaaf, Nijhuis, & Schouten, 2013).

2.6 Industrial upscaling

Upscaling is the migration of a process from the lab-scale to the pilot plant-scale or commercial scale according to scale-up rules. There are always some parameters which are desired to be the same both in pilot and commercial scales. Regarding this fact, the most effective parameters should be selected and their effect must be investigated to migrate from the pilot scale to commercial scale, achieving the same or even better performance. There is a very low number of researches focusing on the upscaling procedures for HiGee equipment.

The main parameters of great importance that should be considered in upscaled versions of ordinary HiGee based sorption equipment are described below.

- Capacity
- Mass Transfer
- Pressure Drop
- Power Consumption

2.6.1 Capacity

The capacity of a HiGee equipment is a function of inside area of the annulus which in turn is proportional to both the inside diameter and the axial length of the packing. This is because the flooding is seen first in that region where both gas and liquid flow per unit area are greatest and "*g*" is smallest. In sum, the capacity of a HiGee equipment has a linear proportionality to both the inside diameter and the axial length of the rotor, while has nonlinear proportionality to the speed of the rotation.

All these points discussed the capacity of a HiGee equipment should be considered in upscaling calculations.

2.6.2 Mass transfer

One of the main important parameters affecting the mass transfer rate is the NTU. In any common HiGee equipment, the mass transfer rate is the function of the distance over the packing the fluids is traversing. In the other words, the NTU is proportional to the difference between the inside and outside radii of the packing. Another important parameter affecting the mass transfer rate is the rotational speed. Generally, increasing the rotational speed would result in the increase in the mass transfer rate. Yet, it should be noted that the rotational speed effectiveness depends on whether the mass resistance is dominant in gas film or liquid film. Despite the advantages of increasing the rotational speed, it brings about the increase in power consumption and pressure drop. As a result, when traveling from a pilot scale to industrial scale, an optimum balance between these conflicting should be considered in the calculation of the rotational speed.

2.6.3 Pressure drop

The size of packing plays a great role in determining the pressure drop either in conventional or HiGee sorption equipment. On the other hand, the volume of packing in HiGee equipment is much smaller than those of conventional equipment. Accordingly, the gas phase pressure drop is of great concern in HiGee equipment. The pressure drop is affected by two factors:

- Friction: as the gas flows through the packing, the generated friction drops the pressure of the gas.
- Rotational speed: the rotational component force the gas periphery to the center of the rotor.

The upscaled and pilot-scale HiGee equipment should not have much difference in the pressure drop value.

2.6.4 Power consumption

The power consumption of the HiGee equipment is rather high, which is mainly due to the rotary parts of them. The larger the equipment, the larger consumption of the energy. So, in industrial scale, it is a concern that must be considered. Three main parameters affecting the power consumption of a HiGee equipment are described in the following. The contribution of each one depends on some parameters like the duty and fluid densities.

- The energy losses including mechanical and electrical and those losses related to frictions between moving parts and the air.
- The consumed energy utilized to accelerate the liquid up to the rotor edge speed.
- Power recovered from the gas traveling to the rotor center.

All four parameters discussed above are believed as essential operating parameter for design and scaling up. These parameters are common. However, there may be some differences for a certain HiGee equipment in selection of the important parameters in upscaling procedure.

2.7 Economic evaluation/justification

Despite the importance of the subject, scarce studies have been performed focusing on the economic evaluation of HiGee equipment. In general, lower investment and operating costs are reported for industrial applications of HiGee equipment demonstrated that generally a major capital cost of an industrial plant including absorption/stripper process are related to absorber and stripper (Abu-Zahra, Schneiders, Niederer, Feron, & Geert, 2007). Indeed, about 55% and 17% of the capital cost are due to the existence of absorber and stripper, respectively. On the other hand, using HiGee technology drastically reduces the size of these equipment. It is therefore expected to lower capital cost by using the HiGee technology. It should be noted that there are many disclosed variables affecting each specific case that must be perfectly investigated and consequently a comprehensive comparison would be performed between the HiGee and conventional sorption equipment in term of economic evaluation.

Economic evaluation related to the HiGee distillation column has been studied by Cortes Garcia et al. (2017). They investigated the main reasons for slow deployment of the HiGee equipment, despite their low capital costs. They listed the main drawbacks, which are often related to reliability and operating costs as additional energy requirements (electricity), complicated hydrodynamics, seals, and bearings among them. In another work by Li (2000) the investment, space, and energy consumption of a conventional tower with HiGee based equipment for the SO_2 adsorption in the acid factory in Zibo of China was compared. It was concluded that HiGee technology decreased from about \$130,000; 120 m^3; and 80 kW, to \$85,000; 60 m^3; and 60 kW, respectively.

Sudhoff et al. (Sudhoff, Leimbrink, Schleinitz, Górak, & Lutze, 2015) studied the investment and operating costs of RPB. They developed some relationships. According to their research, the three groups of costs contribute to the total investment costs of an RBP ($C_{IC\text{-}RBP}$) as follows:

- The costs related to the motor ($C_{IC\text{-}motor}$).
- The costs related to bearings, drives, and rotating parts ($C_{IC\text{-}drive}$).
- The costs related to rotors such as casing, packing, distributors and shafts ($C_{IC\text{-}rotors}$).

Using the factor of C_{tax}, the formulation of $C_{IC\text{-}RBP}$ (in €) is defined as follows:

$$C_{IC,RPB} = \left(C_{IC,mot} + C_{IC,drive} + C_{IC,rotors}\right) C_{tax} \quad (2.5)$$

The operating costs of the RBP include the operating costs of reboiler, condenser, and motor driving the rotor. Following equation can be used for estimation of the operating costs of the motor:

$$C_{OC,\ RPB} = P_{mot} \tau_{hla} C_{el} \tag{2.6}$$

where, P_{mot}, τ_{hla} and C_{el} denote the electrical power of the motor, operating time per year, and the costs for electricity, respectively.

Now the CPT (costs-per-tone value) of a product can be estimated regarding the total investment and operating costs in relation to the amount of product. In addition, the depreciation time of three years as common for small scale investments is taken into account in the formulation:

$$CPT = \left(C_{IC-ot} \tau_{hla}^{-1} + C_{OC-ot}\right)\left(\dot{m}_{dist} \tau_{hla}\right) \tag{2.7}$$

2.8 Future perspectives

In spite of much laboratory-scale research, a limited number of RPB applications at industrial scale were reported. Scale-up of RPB for deoxygenation of well water with capacity of 250 t/h in each unit was successfully performed in the Shengli oil field belonging to China Petrochemical Corporation (Reay et al., 2013). GasTran Systems, a US company, applied RPB for deaeration of water flood for secondary oil recovery on onshore oil wells; based on their report, not only RPB had very good performances but also annual cost was reduced due to prevention of corrosion (Harbold and Park, 2008; Reay et al., 2013). Despite these successfully application, it seems some problems such as sealing requirement and other limitations attributed to the rotating system in comparison with more familiar equipment, such as pumps, compressors, and blowers, are the major problems of industrial scale-up (Hassan-Beck, 1997). Nonetheless, using RPB has very interesting perspectives due to mentioned advantages and overcome to its industrial scale-up limitations is field of many researchers. RPB could see wider applications in future due to its reliability, safety, and efficiency. Commercial experience of RPB such as deaeration and HOCl reactive stripping indicated this technology is very useful, with high performance. Development of theoretical expressions and experimental correlation for more precise prediction of key parameters, such as pressure drop, holdup, residence time, power, and flooding, can be beneficial in design stage. This research, besides commercial applications, help to overcome risks of scale-up.

HiGee technology have found numerous applications in several fields of engineering and science. Some of them are in the first stages of its way, while some have reached maturity. The summary of its applications are:

- Nanoparticles (inorganic, metal, and drug)
- Reaction process intensification (polymerization, HOCl, MDI, sulfonation)
- Separation process intensification (CO_2 capture, devolatilization, deaeration, desulfurization)

Nanoparticle production using HiGee technology is a growing field. Indeed, HiGee technology is utilized to develop the technology related to nanoparticles and nanocomposite. The advantages of application of HiGee technology in this field are producing high contact of inorganic nanoparticle as well as high increased quality of mechanical properties. For instance, the nanocomposites constructed by HiGee technology for silicone sealant increases the tensile strength up to 80% comparing with home-made products (Sudhoff et al., 2015). In future, this application will be paid much attention due to its performance and the growing employments of nanoparticles in numerous fields of engineering and science. The drug nanoparticles are another technology that we will hear more about in the future. Poor water-soluble drugs enhance the solubility and dissolution rates of solid oral formulations. Today, the drug nanoparticles have a great potential of worldwide market and will be increased in the future.

2.9 Opportunities and challenges

Despite many advantages using high-gravity field technology in separation processes like, drastic reduction in the size of equipment, the enhancement of mass transfer and consequently high performance, there are still some challenges acting as a limitation in development in the chemical industries. One of the challenges of HiGee gas-liquid contactor is the amount of their pressure drop which is higher than of conventional packed beds. In addition, there are some problems related to the rotating parts of the HiGee contactors. Indeed, there are concerns about the reliability and longevity of the rotor, bearings, and dynamic seals of the HiGee contactors.

2.10 Conclusion

Designing novel equipment for reduction of capital and operational costs and improvement mass and heat transfer is one the challenges for engineers and scientist in many industries. Process intensification is a strategy which can be used in this field based on miniaturization and decentralization of plant and processes. HiGee is a cutting-edge technology framework for process intensification. HiGee is strongly suggested for industrial innovation of chemical processes limited by low mass transfer rate and micromixing. Engineering of nanoparticles, reaction process intensification, and separation process intensification (especially sorption process) are three main fields have welcome to HiGee technology. This chapter dealt with HiGee sorption technology which reviewed different issues such as mechanism, applications, challenges, and so on. It was discussed that the mechanism of the mass transfer in sorption process under high-gravity has been described by three main theories including film theory, penetration theory, and surface renewal theory. In addition, the effect of different parameters such as the centrifugal force, initial concentration, and the size of adsorbent on the mechanism were investigated. It was concluded that changing the centrifugal field strength can change the pattern of the liquid flow. Furthermore, the smaller size the absorbents, the thinner boundary layer, and faster adsorption.

RPB is one among the most efficient pieces of equipment to practically apply the principles of process intensification, leading to such advantages as cost reduction, efficiency enhancement, and inherent safety. This kind of devices can potentially be applied in many industrial applications, such as absorption, adsorption, stripping, distillation, nanoparticles production, reactive crystallization, polymerization, combustion, precipitation, and competing fast chemical reactions. According to the emphasis of RPB and its applications, fundamental principles of this operation, such as liquid holdup, flooding, RTD, mass transfer, and micromixing, and also applications details for different processes were reviewed in the current chapter.

References

Abu-Zahra, M., Schneiders, L., Niederer, J., Feron, P., & Geert, F. (2007). Versteeg CO_2 capture from power plants: Part II. A parametric study of the economical performance based on mono-ethanolamine. *International Journal of Greenhouse Gas Control, 1*, 135–142.

Becht, S., Franke, R., Geißelmann, A., & Hahn, H. (2009). An industrial view of process intensification. *Chemical Engineering and Processing: Process Intensification, 48*, 329–332.

Boodhoo, K. V. K., & Al-Hengari, S. R. (2012). Micromixing characteristics in a small-scale spinning disk reactor. *Chemical Engineering and Technology, 35*, 1229–1237.

Boodhoo, K. V. K., Vicevic, M., Cartwright, C. D., Ndlovu, T., & Toogood, E. C. (2008). Intensification of gas–liquid mass transfer using a rotating bed of porous packings for application to an *E. coli* batch fermentation process. *Chemical Engineering Journal, 135*, 141–150.

Burns, J. R., Jamil, J. N., & Ramshaw, C. (2000). Process intensification: Operating characteristics of rotating packed beds—determination of liquid hold-up for a high-voidage structured packing. *Chemical Engineering Science, 55*, 2401–2415.

Chang, C.-F., & Lee, S.-C. (2012). Adsorption behavior of pesticide methomyl on activated carbon in a high gravity rotating packed bed reactor. *Water Research, 46*, 2869–2880.

Cheng, H.-H., Lai, C.-C., & Tan, C.-S. (2013). Thermal regeneration of alkanolamine solutions in a rotating packed bed. *International Journal of Greenhouse Gas Control, 16*, 206–216.

Cheng, H.-H., & Tan, C.-S. (2009). Carbon dioxide capture by blended alkanolamines in rotating packed bed. *Energy Procedia, 1*, 925–932.

Cheng, H.-H., & Tan, C.-S. (2011). Removal of CO_2 from indoor air by alkanolamine in a rotating packed bed. *Separation and Purification Technology, 82*, 156–166.

Chen, Y.-S., Hsu, Y.-C., Lin, C.-C., Tai, C. Y. D., & Liu, H.-S. (2008). Volatile organic compounds absorption in a cross-flow rotating packed bed. *Environmental Science and Technology, 42*, 2631–2636.

Chen, Y.-S., Lin, C.-C., & Liu, H.-S. (2005). Mass transfer in a rotating packed bed with various radii of the bed. *Industrial & Engineering Chemistry Research, 44*, 7868–7875.

Chen, Y.-S., & Liu, H.-S. (2002). Absorption of VOCs in a rotating packed bed. *Industrial and Engineering Chemistry Research, 41*, 1583–1588.

Chiu, C.-Y., Chen, Y.-H., & Huang, Y.-H. (2007). Removal of naphthalene in Brij 30-containing solution by ozonation using rotating packed bed. *Journal of Hazardous Materials, 147*, 732–737.

Cortes Garcia, G. E., van der Schaaf, J., & Kiss, A. A. (2017). A review on process intensification in HiGee distillation. *Journal of Chemical Technology and Biotechnology, 92*, 1136–1156.

Das, A., Bhowal, A., & Datta, S. (2008). Continuous biosorption in rotating packed-bed contactor. *Industrial & Engineering Chemistry Research, 47*, 4230–4235.

Ehinmowo, Bolu, A., Aliyu, A. M., Alagbe, S. O., Adedipe, O., & Olumi, O. (2015). Offshore topside rotating packed bed as process intensified alternative for natural gas sweetening and dehydration. *Chemical Science International Journal.*

Guo, K., Guo, F., Feng, Y., Chen, J., Zheng, C., & Gardner, N. C. (2000). Synchronous visual and RTD study on liquid flow in rotating packed-bed contactor. *Chemical Engineering Science, 55*, 1699–1706.

Guo, K., Zhang, Z., Luo, H., Dang, J., & Qian, Z. (2014). An innovative approach of the effective mass transfer area in the rotating packed bed. *Industrial and Engineering Chemistry Research, 53*, 4052–4058.

Guo, F., Zheng, C., Guo, K., Feng, Y., & Gardner, N. C. (1997). Hydrodynamics and mass transfer in cross-flow rotating packed bed. *Chemical Engineering Science, 52*, 3853–3859.

Hao, S., & Nangeng, S. (1994). Stripping ammonia from ammonia water IN centrifugal mass transfer apparatus. *Journal of Nanjing University of Chemical Technology, 4.*

Harbold, G., & Park, J. (2008). *Using the GasTran® deaeration system to achieve low dissolved oxygen levels for superior line speed and product quality: A case study in carbonated soft drink bottling.*

Hassan-Beck, H. M. (1997). *Process intensification: Mass transfer and pressure drop for countercurrent rotating packed beds.*

Hsu, L.-J., & Lin, C.-C. (2011). Removal of methanol and 1-butanol from binary mixtures by absorption in rotating packed beds with blade packings. *Chemical Engineering Journal, 168*, 190–200.

Jachuck, R. J., Lee, J., Kolokotsa, D., Ramshaw, C., Valachis, P., & Yanniotis, S. (1997). Process intensification for energy saving. *Applied Thermal Engineering, 17*, 861–867.

Jiang, X., Liu, Y., & Gu, M. (2011). Absorption of sulphur dioxide with sodium citrate buffer solution in a rotating packed bed. *Chinese Journal of Chemical Engineering, 19*, 687–692.

Jiao, W., Liu, Y., & Qi, G. (2010b). A new impinging stream–rotating packed bed reactor for improvement of micromixing iodide and iodate. *Chemical Engineering Journal, 157*, 168–173.

Jiao, W. Z., Liu, Y. Z., & Qi, G. S. (2010a). Gas pressure drop and mass transfer characteristics in a cross-flow rotating packed bed with porous plate packing. *Industrial and Engineering Chemistry Research, 49*, 3732–3740.

Joel, A. S., Wang, M., & Ramshaw, C. (2015). Modelling and simulation of intensified absorber for post-combustion CO_2 capture using different mass transfer correlations. *Applied Thermal Engineering, 74*, 47–53.

Kang, J.-L., Luo, Z.-J., Liu, J.-L., Sun, K., Hill Wong, D. S., Jang, S.-S., et al. (2014). Experiment and modeling studies on absorption of CO_2 by dilute ammonia in rotating packed bed. *Energy Procedia, 63*, 1308–1313.

Kang, J.-L., Sun, K., Wong, D. S.-H., Jang, S.-S., & Tan, C.-S. (2014). Modeling studies on absorption of CO_2 by monoethanolamine in rotating packed bed. *International Journal of Greenhouse Gas Control, 25*, 141–150.

Keyvani, M., & Gardner, N. C. (1989). Operating characteristics of rotating beds. *Chemical Engineering Progress, 85*, 48–52.

Kumar, M. P., & Rao, D. P. (1990). Studies on a high-gravity gas-liquid contactor. *Industrial & Engineering Chemistry Research, 29*, 917–920.

Kundu, A., SenGupta, B., Hashim, M. A., & Redzwan, G. (2015). Taguchi optimisation approach for chromium removal in a rotating packed bed contractor. *Journal of the Taiwan Institute of Chemical Engineers, 57*, 91–97.

Li, Z. H. (2000). A study on modeling of mass transfer in RPB. In *Beijing University of chemical technology, China.*

Lin, C.-C., & Chen, B. C. (2008). Characteristics of cross-flow rotating packed beds. *Journal of Industrial and Engineering Chemistry, 14*, 322–327.

Lin, C.-C., Chao, C.-Y., & Liu, M.-Y. (2010). Removal of ozone from air by absorption in a rotating packed bed. *Journal of Industrial and Engineering Chemistry, 16*, 140–146.

Lin, C.-C., Chao, C.-Y., Liu, M.-Y., & Lee, Y.-L. (2009). Feasibility of ozone absorption by H_2O_2 solution in rotating packed beds. *Journal of Hazardous Materials, 167*, 1014–1020.

Lin, C.-C., & Chen, Y.-W. (2011a). Carbon dioxide absorption in a cross-flow rotating packed bed. *Chemical Engineering Research and Design, 89*, 1722–1729.

Lin, C.-C., & Chen, Y.-W. (2011b). Performance of a cross-flow rotating packed bed in removing carbon dioxide from gaseous streams by chemical absorption. *International Journal of Greenhouse Gas Control, 5*, 668–675.

Lin, C.-C., Chen, Y.-S., & Liu, H.-S. (2000). Prediction of liquid holdup in countercurrent-flow rotating packed bed. *Chemical Engineering Research and Design, 78*, 397–403.

Lin, C.-C., & Chien, K.-S. (2008). Mass-transfer performance of rotating packed beds equipped with blade packings in VOCs absorption into water. *Separation and Purification Technology, 63*, 138–144.

Lin, C.-C., & Chu, C.-R. (2015). Feasibility of carbon dioxide absorption by NaOH solution in a rotating packed bed with blade packings. *International Journal of Greenhouse Gas Control, 42*, 117–123.

Lin, C.-C., & Jian, G.-S. (2007). Characteristics of a rotating packed bed equipped with blade packings. *Separation and Purification Technology, 54*, 51–60.

Lin, C.-C., Lin, Y.-C., & Chien, K.-S. (2009). VOCs absorption in rotating packed beds equipped with blade packings. *Journal of Industrial and Engineering Chemistry, 15*, 813–818.

Lin, C.-C., & Liu, W.-T. (2007). Mass transfer characteristics of a high-voidage rotating packed bed. *Journal of Industrial and Engineering Chemistry, 13*, 71–78.

Lin, C.-C., Wei, T.-Y., Hsu, S.-K., & Liu, W.-T. (2006). Performance of a pilot-scale cross-flow rotating packed bed in removing VOCs from waste gas streams. *Separation and Purification Technology, 52*, 274–279.

Lin, C.-C., Wei, T.-Y., Liu, W.-T., & Shen, K.-P. (2004). Removal of VOCs from gaseous streams in a high-voidage rotating packed bed. *Journal of Chemical Engineering of Japan, 37*, 1471–1477.

Liu, H.-S., Lin, C.-C., Wu, S.-C., & Hsu, H.-W. (1996). Characteristics of a rotating packed bed. *Industrial and Engineering Chemistry Research, 35*, 3590–3596.

Li, Y., YuLi, Y., XuLi, Z., LiLi, X., Liu, X., & Ji, J. (2013). Rotating zigzag bed as trayed HIGEE and its power consumption. *Asia-Pacific Journal of Chemical Engineering, 8*, 494–506.

Llerena-Chavez, H., & Larachi, F. (2009). Analysis of flow in rotating packed beds via CFD simulations—dry pressure drop and gas flow maldistribution. *Chemical Engineering Science, 64*, 2113–2126.

Luo, Y., Chu, G.-W., Zou, H.-K., Zhao, Z.-Q., Dudukovic, M. P., & Chen, J.-F. (2012). Gas–liquid effective interfacial area in a rotating packed bed. *Industrial and Engineering Chemistry Research, 51*, 16320–16325.

Martínez, E. L., Jaimes, R., Gomez, J. L., & Filho, R. M. (2012). CFD simulation of three-dimensional multiphase flow in a rotating packed bed. In B. I. D. Lockhart, & F. Michael (Eds.), *Computer aided chemical engineering*. Elsevier.

Meeuwse, M., van der Schaaf, J., & Jaap, C. S. (2010). Mass transfer in a rotor– Stator spinning disk reactor with cofeeding of gas and liquid. *Industrial and Engineering Chemistry Research, 49*, 1605–1610.

Mosleh, S., & Rahimi, M. R. (2016). Intensification of Abamectin pesticide degradation using the combination of ultrasonic cavitation and visible-light driven photocatalytic process: Synergistic effect and optimization study. *Ultrasonics Sonochemistry*.

Mosleh, S., Rahimi, M. R., Ghaedi, M., & Dashtian, K. (2016). Sonophotocatalytic degradation of trypan blue and vesuvine dyes in the presence of blue light active photocatalyst of Ag_3PO_4/Bi_2S_3-HKUST-1-MOF: Central composite optimization and synergistic effect study. *Ultrasonics Sonochemistry, 32*, 387–397.

Mosleh, S., Rahimi, M. R., Ghaedi, M., Dashtian, K., & Hajati, S. (2016a). $BiPO_4/Bi_2S_3$-HKUST-1-MOF as a novel blue light-driven photocatalyst for simultaneous degradation of toluidine blue and auramine-O dyes in a new rotating packed bed reactor: Optimization and comparison to a conventional reactor. *RSC Advances, 6*, 63667–63680.

Mosleh, S., Rahimi, M. R., Ghaedi, M., Dashtian, K., & Hajati, S. (2016b). Photocatalytic degradation of binary mixture of toxic dyes by HKUST-1 MOF and HKUST-1-SBA-15 in a rotating packed bed reactor under blue LED illumination: Central composite design optimization. *RSC Advances, 6*, 17204–17214.

Mosleh, S., Rahimi, M. R., Ghaedi, M., Dashtian, K., Hajati, S., & Wang, S. (2017). Ag_3PO_4/AgBr/Ag-HKUST-1-MOF composites as novel blue LED light active photocatalyst for enhanced degradation of ternary mixture of dyes in a rotating packed bed reactor. *Chemical Engineering and Processing: Process Intensification*.

Munjal, S., Dudukovć, M. P., & Ramachandran, P. (1989a). Mass-transfer in rotating packed beds—I. Development of gas—liquid and liquid—solid mass-transfer correlations. *Chemical Engineering Science, 44*, 2245–2256.

Munjal, S., Duduković, M. P., & Ramachandran, P. (1989b). Mass-transfer in rotating packed beds—II. Experimental results and comparison with theory and gravity flow. *Chemical Engineering Science, 44*, 2257–2268.

Nikačević, N. M., Huesman, A. E. M., Van den Hof, P. M. J., & Stankiewicz, A. I. (2012). Opportunities and challenges for process control in process intensification. *Chemical Engineering and Processing: Process Intensification, 52*, 1–15.

Pan, S.-Y., Chiang, P.-C., Chen, Y.-H., Tan, C.-S., & Chang, E.-E. (2014). Kinetics of carbonation reaction of basic oxygen furnace slags in a rotating packed bed using the surface coverage model: Maximization of carbonation conversion. *Applied Energy, 113*, 267–276.

PMNT. (2004). Processing. In *Re-engineering*.

Ponce-Ortega, J. M., Al-Thubaiti, M. M., & El-Halwagi, M. M. (2012). Process intensification: New understanding and systematic approach. *Chemical Engineering and Processing: Process Intensification, 53*, 63–75.

Qian, Z., Chen, Q., & Grossmann, I. E. (2018). Optimal synthesis of rotating packed bed and packed bed: A case illustrating the integration of PI and PSE. In *Computer aided chemical engineering*. Elsevier.

Rahimi, M. R., Karimi, H., & Mosleh, S. (2013). Experimental study of carbon dioxide absorption from air stream in rotating packed bed. *Journal of Separation Science and Engineering, 5*, 1–16.

Rahimi, M. R., & Mosleh, S. (2013). Mass transfer modeling for volatile organic compounds absorption in rotating packed beds. *Journal of Separation Science and Engineering, 5*, 1–10.

Rahimi, M. R., & Mosleh, S. (2015). CO_2 removal from air in a countercurrent rotating packed bed, experimental determination of height of transfer unit. *Advances in Environmental Science and Technology, 1*, 19–24.

Rahimi, M. R., & Zarei, F. (2016). Experimental study of overall gas phase mass transfer coefficient and height of transfer unit in countercurrent rotating packed bed. *Journal of Separation Science and Engineering, 8*, 45–56.

Ramshaw, C. (1993). *HeatRecoVery Syst. 13* p. 493). CHP.

Rao, D. P. (2015). The story of "HIGEE". *Indian Chemical Engineer, 57*, 282–299.

Reay, D., Ramshaw, C., & Harvey, A. (2013). *Process Intensification: Engineering for efficiency, sustainability and flexibility*. Butterworth-Heinemann.

Sandilya, P., Rao, D. P., Sharma, A., & Biswas, G. (2001). Gas-phase mass transfer in a centrifugal contactor. *Industrial & Engineering Chemistry Research, 40*, 384–392.

Shivhare, M. K., Rao, D. P., & Kaistha, N. (2013). Mass transfer studies on split-packing and single-block packing rotating packed beds. *Chemical Engineering and Processing: Process Intensification, 71*, 115–124.

Shi, X., Yang, X., Wen, L.-X., & Chen, J.-F. (2013). CFD analysis of liquid phase flow in a rotating packed bed reactor. *Chemical Engineering Journal, 228*, 1040–1049.

Stakić, M., & Urošević, T. (2011). Experimental study and simulation of vibrated fluidized bed drying. *Chemical Engineering and Processing: Process Intensification, 50*, 428–437.

Stefanidis, G. D., & Stankiewicz, A. (2013). Process Intensification by alternative energy forms and transfer mechanisms. *Chemical Engineering and Processing, 71*, 1–1.

Sudhoff, D., Leimbrink, M., Schleinitz, M., Górak, A., & Lutze, P. (2015). Modelling, design and flexibility analysis of rotating packed beds for distillation. *Chemical Engineering Research and Design, 94*, 72–89.

Sui, L., Xu, Z., Yu, Y., & Ji, J. (2008). Influence of the rotor structure of rotating zigzag bed on gas pressure drop. *Journal of Chemical Engineering of Chinese Universities, 22*, 28.

Sung, W.-D., & Chen, Y.-S. (2012). Characteristics of a rotating packed bed equipped with blade packings and baffles. *Separation and Purification Technology, 93*, 52–58.

Tan, C.-S., & Lee, P.-L. (2008). Supercritical CO_2 desorption of toluene from activated carbon in rotating packed bed. *The Journal of Supercritical Fluids, 46*, 99–104.

Visscher, F., Van der Schaaf, J., Nijhuis, T. A., & Schouten, J. C. (2013). Rotating reactors—a review. *Chemical Engineering Research and Design, 91*, 1923–1940.

Wang, G. Q., Xu, Z. C., Yu, Y. L., & Ji, J. B. (2008). Performance of a rotating zigzag bed—a new HIGEE. *Chemical Engineering and Processing: Process Intensification, 47*, 2131–2139.

Woyuan, L., Wu, W., Zou, H., Chu, G., Lei, S., & Chen, J. (2009). Process intensification of VOC removal from high viscous media by rotating packed bed. *Chinese Journal of Chemical Engineering, 17*, 389–393.

Xie, P., Lu, X., Ding, H., Yang, X., Derek Ingham, Lin, M., et al. (2019). A mesoscale 3D CFD analysis of the liquid flow in a rotating packed bed. *Chemical Engineering Science, 199*, 528–545.

Xiuping, L., & Youzhi, L. (2010). Characteristics of fin baffle packing used in rotating packed bed. *Chinese Journal of Chemical Engineering, 18*, 55–60.

Yang, K., Chu, G.-W., Shao, L., Luo, Y., & Chen, J.-F. (2009). Micromixing efficiency of rotating packed bed with premixed liquid distributor. *Chemical Engineering Journal, 153*, 222–226.

Yang, H. J., Chu, G. W., Yang, X., & Chen, J. F. (2006). Characterization of micromixing efficiency in rotating packed beds by chemical methods. *Chemical Engineering Journal, 121*, 147–152.

Yang, K., Chu, G., Zou, H., Sun, B., Shao, L., & Chen, J.-F. (2011). Determination of the effective interfacial area in rotating packed bed. *Chemical Engineering Journal, 168*, 1377–1382.

Yang, W., Wang, Y., Chen, J., & Fei, W. (2010). Computational fluid dynamic simulation of fluid flow in a rotating packed bed. *Chemical Engineering Journal, 156*, 582–587.

Yi, F., Zou, H.-K., Chu, G.-W., Shao, L., & Chen, J.-F. (2009). Modeling and experimental studies on absorption of CO_2 by Benfield solution in rotating packed bed. *Chemical Engineering Journal, 145*, 377–384.

Yu, C.-H., Cheng, H.-H., & Tan, C.-S. (2012). CO_2 capture by alkanolamine solutions containing diethylenetriamine and piperazine in a rotating packed bed. *International Journal of Greenhouse Gas Control, 9*, 136–147.

Yu, L., Liu, Y., Zhang, L., Su, Q., & Jin, G. (2010). Absorption of NO_x into nitric acid solution in rotating packed bed. *Chinese Journal of Chemical Engineering, 18*, 244–248.

Yu, C.-H., & Tan, C.-S. (2014). CO_2 capture by aqueous solution containing mixed alkanolamines and diethylene glycol in a rotating packed bed. *Energy Procedia, 63*, 758–764.

Zhang, L.-L., Wang, J.-X., Sun, Q., Zeng, X.-F., & Chen, J.-F. (2012). Removal of nitric oxide in rotating packed bed by ferrous chelate solution. *Chemical Engineering Journal, 181*, 624–629.

Zhang, W., Xie, P., Li, Y., Teng, L., & Zhu, J. (2020). Hydrodynamic characteristics and mass transfer performance of rotating packed bed for CO_2 removal by chemical absorption: A review. *Journal of Natural Gas Science and Engineering*, 103373.

Zhao, H., Shao, L., & Chen, J.-F. (2010). High-gravity process intensification technology and application. *Chemical Engineering Journal, 156*, 588–593.

Zhao, B., Su, Y., & Tao, W. (2014). Mass transfer performance of CO_2 capture in rotating packed bed: Dimensionless modeling and intelligent prediction. *Applied Energy, 136*, 132–142.

CHAPTER THREE

Magnetic and electrical-assisted adsorption processes

3.1 Introduction

The adsorption procedures for separation purposes involve a porous solid and a fluid mix (gas or liquid, which only the usage of gas phase is taken into account) (Liu, Chen, & Gao, 2017). The entire cycle of the separation process consists of two half-cycles, which are as follows: a cold half-cycle and a hot half-cycle for adsorption and desorption, respectively. In the cold half-cycle, the gas mix passes through sorbent bed to be separated which the target particles were absorbed, which is obtained by outlet from a depleted amount of particles (Otero, Zabkova, Grande, & Rodrigues, 2005). When the adsorption capacity (saturation) is achieved, the hot half-cycle starts to regenerating absorbent with a reflux or purge gas. Adsorption and desorption are carried out one after the other, causing the separation of the gas mixture.

The cyclical characteristics of the separation processes are intensely fundamental due to the determination of the cycle duration and productivity (Thouchprasitchai, Pintuyothin, & Pongstabodee, 2018). Consequently, the key purpose is to shorten the cycle duration for intensifying the procedure. Here the aspect of "temporal intensification" has been considered, while geometric or spatial intensification aspects are also important (Li, Wu, Hao, Li, & Gao, 2019). Dependence to the mass and heat transfer properties is an important feature of adsorption processes. At low temperatures, adsorption increases, while at high temperatures, desorption is improved. Therefore, the productivity of such procedure is heavily affected by the frequency that we can change the sorbent bed temperature, i.e., heating or cooling the sorbent for reaching to the set point of adsorption and desorption over shortest possible time (Qasem & Ben-Mansour, 2018).

However, the internal structure of porous media is one of the most influential factors in adsorption/desorption duration, for example, the transfer time for passing molecules of gas through meso- and micropores, and eventually reaching the nano pores where adsorption occurs (Tang et al., 2020).

Intensification of Sorption Processes
ISBN: 978-0-12-821411-4
https://doi.org/10.1016/B978-0-12-821411-4.00005-5

Porous media porosity must also be obtained correctly, since more solid leads to higher heat conductivity, while resulting to less pore volume for adsorption capacity. Consequently, how to design a porous material that provides the optimum distribution of the pores volume at diverse scales to achieve the simplest and most uniform gas molecules toward nano pores is a major problem in developing the productivity of the adsorption process (Qu, Yin, Wang, & Zhang, 2020). In this chapter, efforts have been carried out to demonstrate intensifying of the sorption processes using some effective techniques such as thermal swing adsorption, electrothermal swing adsorption, pressure swing adsorption, electrohydraulic discharge, magnetic adsorption, high gradient magnetic separation and magnetic induction swing adsorption.

3.2 Thermal swing adsorption (TSA)

3.2.1 TSA fundamentals

The TSA is utilized to purify gas molecules like volatile organic compounds (VOCs) treatment from gaseous streams (Fig. 3.1). In conventional TSA procedures, heating and adsorption typically influenced by steam or hot gas via the bed, and is used to cool cold gas (Zhao et al., 2019).

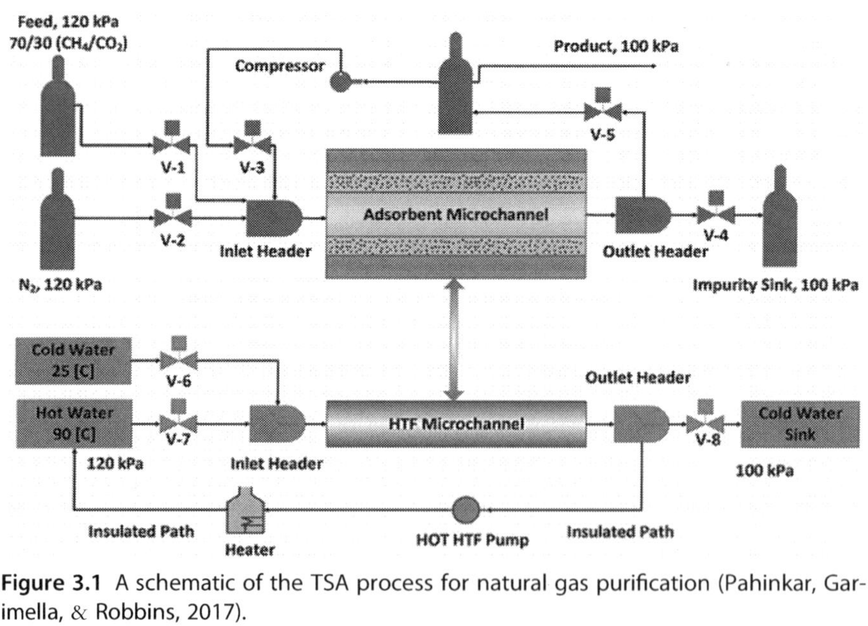

Figure 3.1 A schematic of the TSA process for natural gas purification (Pahinkar, Garimella, & Robbins, 2017).

Steam adsorption is commonly fast and effective, but needs two steps of drying and cooling. In all cases, this means that a large amount of gas is needed to pass through the bed and it usually takes a few hours (Jiang, Roskilly, & Wang, 2018).

This section discusses about the different intensification methods that utilized, containing:

- Increasing the total heat transfer in the particle bed with "thermal conductivity promoters," e.g., compacted natural graphite sheets;
- Using thermoelectric elements (Peltier effect) for accelerating the temperature change dynamics;
- Using a proper operation mode for the adsorption and desorption cycle;
- Increased separation using cycling-zone adsorption.

The main parameter limiting the time of a TSA cycle is the inappropriate heat transfer, which is directly related to the weak thermal conductivity of the sorbent. As an example, the thermal conductivity of activated carbon is usually about 0.2 $Wm^{-1}K^{-1}$, which can be taken into account as a nonconductive material. While the thermal conductivity of the graphite (for example, compacted natural exfoliated graphite) is much more up to 350W/m K (Bonnissel, Luo, & Tondeur, 2001).

3.2.2 Thermoelectric elements

After describing the sorbent bed shape with increased heat transfer properties, another problem that needs to be taken into account to shorten the cycle, is the mode of supply/discharge of energy, because as we know, adsorption is exothermic and heat is released over this process, though desorption is endothermic and heat is adsorbed over it (Luo, 2013). Typically, the working gas is preheated or cooled prior entering to the sorbent bed. Other approaches are the integration of tubular heat exchangers or the using Joule effect.

3.2.3 TSA processes

The four operation phases include the basic two-bed thermal swing procedure (Fig. 3.2). The components are separated in the first phase (adsorption) of the thermal swing cycle, the component that adsorbed most toughly is retained in the sorbent bed whereas the component that adsorbed least toughly passes through the adsorbent bed. Therefore, the feed mix comprising an adsorbate at relative pressure passes through the sorbent bed, and operate at specific temperature, which the adsorbate completely or incompletely eliminated from the feed.

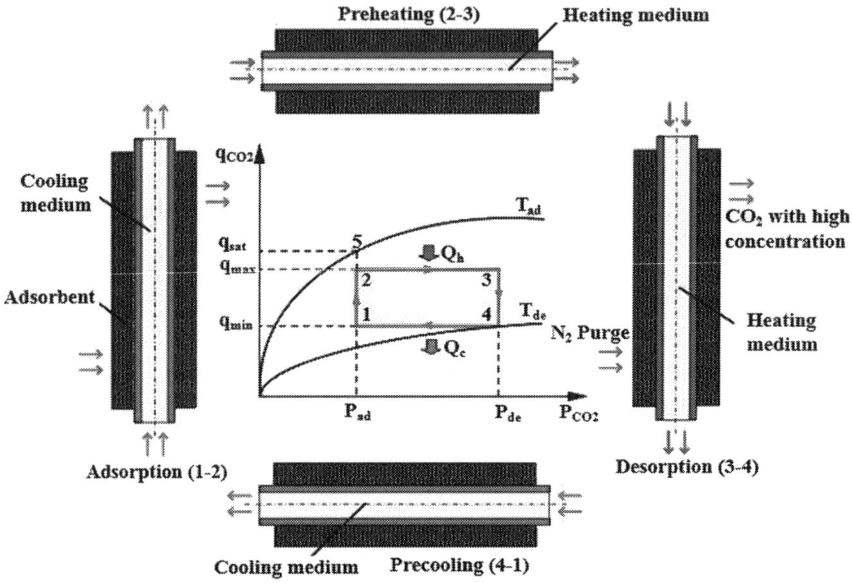

Figure 3.2 A schematic of 4-step TSA process for CO_2 capture (Jiang et al., 2020).

In a two-bed thermal swing procedure, the restriction caused by the equal time of adsorption and desorption reduces the process flexibility and decreases the effective loading capacity corresponding to the beds (Jiang, Liang et al., 2018). If the most toughly adsorbed component in the mix is isolated, the equilibrium isotherm is generated in the axis, which indicates the amount of adsorbed (the desired isotherm mode), the desorption step of the cycle becomes limitation parameter of the total cycle time. In the case of an adsorption isotherm with desired mode, the concentration wave that passes over the bed tends to a constant pattern, however, over desorption, the wave is dispersed, and therefore extends the mass transfer zone (MTZ). A longer period is needed for adsorption and desorption if both the flow rate and temperature are kept at the similar values for adsorption and desorption (Lei, Vallières, Grévillot, & Latifi, 2013). Therefore, the condition for same time of adsorption and desorption in the two-bed temperature swing cycle shows that only the adsorbate fraction present in the feed can be eliminated by adsorption cycle. As a result, the capacity of bed is not completely used. Due to the long cycle times required to the thermal swing separation, this type of operation is utilized nearly exclusively to remove the low concentrations of absorbable gases from feed streams (Joss, Gazzani, & Mazzotti, 2017). In addition, substantial values of energy can be utilized to supply sensible heat and desorption heat, unless the concentration of the removable component is low.

3.2.4 TSA commercial processes

Commercial application of the thermal swing process is for purifying gases and liquids, rather than to separate the bulk gases. Examples include drying air or natural gas (Wurzbacher, Gebald, Piatkowski, & Steinfeld, 2012), removing solvent vapors from the air flow (Ambrożek & Zwarycz-Makles, 2014), sweetening natural gas (Pahinkar, Garimella, & Robbins, 2017), removing ethyl acetate from aromatics and purifying liquid organic compounds (Ajenifuja, Joss, & Jobson, 2020). An example for separating the bulk of the components by adsorbing TSA is extracting water from ethanol (Hanchate, Kulshreshtha, & Mathpati, 2019).

3.2.4.1 Drying gases

Removing trace amounts of water vapor from gases or air is critical to protecting compressors and electronic equipment. Dew point may be attained in the range of -30 to $-50°C$. Therefore, for drying the gas, it initially passes through an adsorption bed, where moisture is eliminated with appropriate porous materials for drying (Kannan, Arjunan, & Vijayan, 2020). While a bed adsorbs moisture, and the other bed is reformed at a higher temperature. When the regeneration is completed, the operation returns to normal, now the second bed acts as an adsorbent though the first bed is regenerating (Ge, Wu, Wang, & Wu, 2019). The selection of sorbent for the bed is affected by the specific application of drying. If the average humidity is needed for air or process gas, the silica gel is absorbent because of its high capacity and easy regeneration. On the other hand, if materials with small dew point are needed, next a molecular sieve is preferable.

3.2.4.2 Gas sweetening

Natural gas stream occasionally contains components including hydrogen sulfide, mercaptane, carbon dioxide and water vapor (Aleghafouri & Davoudi, 2018; Raganati, Chirone, & Ammendola, 2020). To eliminate these unwanted natural gas components (considered as sweetening in the petrochemical industry), a three-bed system can be used. A zeolite with great pores is utilized as the sorbent, the adsorption capacity of the removed components is $H_2O > H_2S > CO_2$, respectively.

3.2.4.3 Removing water from volatile organic components

Based on the nature of the procedure, an organic component comprising water may be a vapor or liquid stream. If, in the formation of noncondensed steam, as an example, water should be removed from a distillation column by means of two adsorption columns, each column will alternately be used as an adsorbent and a regenerator. In addition, with a wise selection of working conditions, significant amounts of adsorption heat can be maintained in the adsorbent bed to ease easy recovery. If the volatile organic mix and the water are in liquid form, a system with two columns may also be utilized, however there are no advantages in maintaining the heat in the adsorption bed (Benson & George, 2005). Two examples are listed for removing water from an alcohol-water mix with composition close to the azeotrope. The first example is removing water from a steam stream, and the second one is the liquid stream. The typical absorbent utilized in both steam and liquid process is a molecular sieve of zeolite. A special zeolite utilized to extract water from a liquid stream or ethanol-water vapor is type 3A, in which sodium is replaced with potassium ion exchange (Hanchate et al., 2019). The molecular dimension of water is 0.28 nm, so water molecules can be adsorbed by the sieve, but the sieve keeps out the larger molecules of ethanol (0.44 nm).

3.3 Electrothermal swing adsorption (ESA)

The ESA concept was originally introduced by scientists at the Oak Ridge National Laboratory. This procedure still involves an adsorption procedure followed by a desorption procedure, such as the traditional TSA or PSA procedure (Fig. 3.3). The unique property is that, at the desorption step, reformation is accomplished by enhancing the adsorbent temperature by using the Joule effect corresponding to passing electricity from an electrically saturated conductive sorbent (Joule's first law of heating) (Johnsen, Emamipour, Guest, & Rood, 2016). Actually, adsorbent is regenerated via reduction of the material's equilibrium capacity. In these circumstances, the principles of operating the ESA procedure appear to be as same as the TSA's performance (Zhao, Liu, Zhao, Deng, & Li, 2018). The in situ heat pathway can efficiently heat sorbents with no heating the additional environment, so the ESA has some benefits, compared to PSA and TSA, such as less heat demand, high heating rates, better kinetics and dynamic of desorption, and independent controlling the gas and heat flow rates.

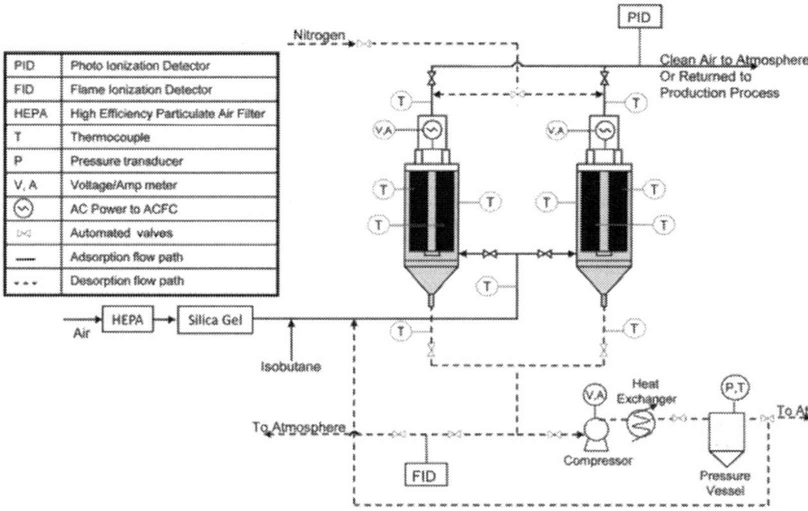

Figure 3.3 A schematic of ESA process for capture and recovery of isobutane (Mallouk, Johnsen, & Rood, 2010).

Alternatively, the sorbent used in ESA experiments has great temperature differences, so there are major differences in loading at low and high temperatures that commonly referred to great values of adsorption heat, and highest efficiency can be attained when low amount of gas is adsorbed (Zhao, Liu, et al., 2018). This is the reason for unsuitability of this procedure for removing large quantities of gas.

When an electric current flows over a finite conductive solid, the electrical energy is changed to heat over resistive losses in the material. Heat is produced on a microscopic scale by transferring energy to the conductor atoms with conduction electrons through collisions. The heat energy released in a conductor during passage of an electric current is known as the "Joule heat" and the effect is known as "Joule effect". Joule's first law of heating states that "the heat produced per unit time, is proportional to the square of the electric current, on passing an electrical current from a conductor at a given temperature."

In a column, the Joule effect can be achieved in two diverse methods. The simplest one is to use a sorbent that can be conduct the electricity (direct ESA) (Emamipour et al., 2007). Using a nonconductive adsorbent and placing some devices in a column that can conduct electricity can be an alternative (indirect ESA) (Ried, Salazar Duarte, & Hinrichsen, 2020).

3.4 Pressure swing adsorption

Many advanced PSA procedures and designs have been applied to intensification of various separation processes (Fig. 3.4). Although, there are not enough basic ideas to made these designs. These ideas can be slightly modified, so that the separation can be done (Tao, Xiao, Qader, & Webley, 2019). To facilitate the below discussions, some terms are defined for the first time. The efficiency of separation procedure is determined on the basis of three factors: (1) purity of product, (2) recovery of product, and (3) productivity of sorbent. In the case of regenerating more than one component, (1) and (2) all products are applied. The purity of the product is a volume-averaged quantity, as the output concentration and flow rate from a PSA procedure change with time, without using surge or combining tanks.

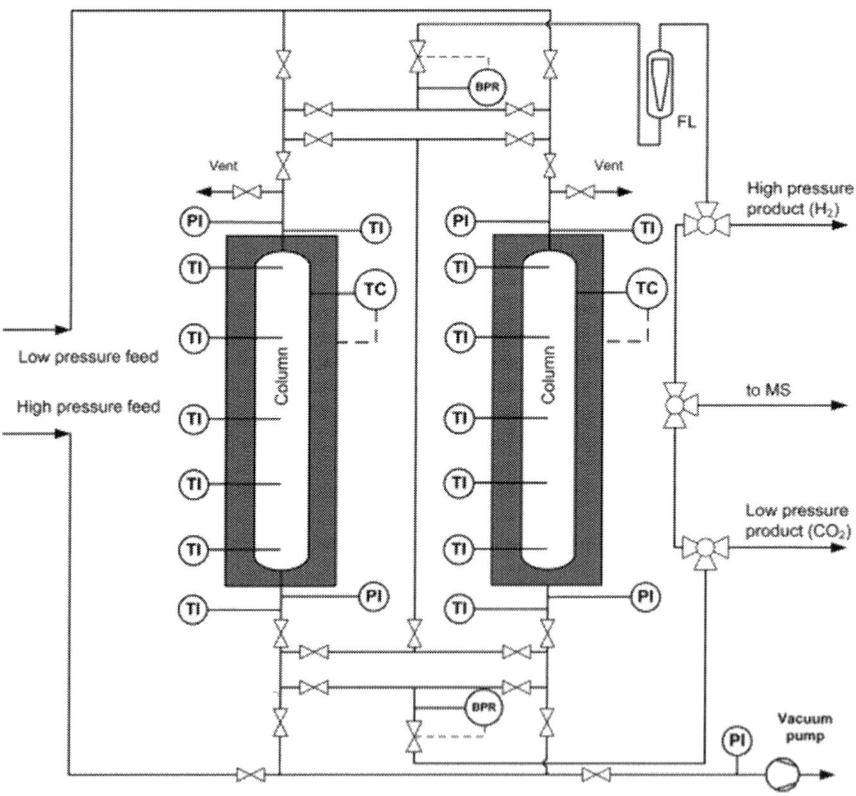

Figure 3.4 A laboratory two-column PSA setup for precombustion CO_2 Capture (Schell, Casas, Marx, & Mazzotti, 2013).

Recovery of product determines the quantity of material in the product stream divided by the quantity of the similar component in the feed mix. Productivity of adsorption is determined using the quantity of product or mixture of processed feed per unit of adsorbent amount per unit of time. For particular separation, the purity of the product is preset, the required energy is typically proportional to recovery, and the inverse of the adsorbent bed size is related proportionally to the adsorbent productivity (Wiheeb, Helwani, Kim, & Othman, 2016). It should be noted that these three parameters are related to each other for any specific PSA process and their relationship is measured only by a model for the procedure (Zhao, Deng, et al., 2018). Besides these factors, the term "strong adsorptive" is utilized to refer to strongly adsorbable gas, while "weak adsorptive" is utilized for poorly adsorbable gas. A product that contains frequently weak adsorptive can also be named "loosely light product" or "raffinate"; the latter is similar to liquid-liquid extraction.

3.4.1 Current improvements and future prospects for PSA

Using the PSA as a tool for enriching gas has only recently been examined. Several investigations have been performed on the purification of air and concentration in a wide range of adsorbents, such as CO_2, SO_2 and gasoline vapors (Abdullah, Idris, Shamsudin, & Othman, 2019; Siqueira et al., 2017). Initial PSA cycles normally included: (1) pressurization of feed; (2) high pressure feed; (3) countercurrent evacuation; and (4) a countercurrent vacuum purge. Vacuuming over desorption is very essential for enrichment, but high pressure is not essential. For the most enrichment, the desired purge/feed ratio is required. This kind of process has been developed at the Korean Energy Research Institute. For enrichment of Xe from air, a two steps PSA was utilized using two notions for obtaining enrichment value of 4000 (from 0.1 ppm in feed to 400 ppm in desorption product) (Banerjee, Simon, Elsaidi, Haranczyk, & Thallapally, 2018). The two main notions were: reflux (or washing) and parallel PE (as described above). The first PSA is ultraatmospheric and the second is under atmospheric condition, both comprising 13X zeolites (Delgado et al., 2014). Considering that ecological uses are becoming growingly important for PSA, further investigations are required on adsorbent enrichment and efficiency. As have observed earlier, the basic and operational problems for enrichment are not as same as those that are involved in purifying and separating the bulk, and are not well known.

3.5 Electrohydraulic discharge process

The usage of strong electric fields in water and organic liquids has been examined for a long time, due to its important role in electrical transfer procedures and its practical application in biology, chemistry and electrochemical (Mouele, Tijani, Fatoba, & Petrik, 2015; Ren et al., 2019). In recent years, liquid-phase electrical discharge reactors have been explored and are being improved for numerous ecological applications, such as drinking water and wastewater purification, as well as possibly environmental safe chemical methods (Jiang et al., 2014). In general, strong electric fields for water (electrohydraulic discharge) starts both chemical and physical procedures (Fig. 3.5).

The two main types of electrohydraulic discharges are primarily defined by the quantity of energy stored in the system. The corona or corona like system utilizes a ~1 j/pulse discharge, while the pulsed arc discharge uses ~1 kJ/pulse and greater (Locke, Sato, Sunka, Hoffmann, & Chang, 2006). The pulse corona system works at 102−103 Hz and peak current is < 100 A, with a voltage increase of 1 ns. A pseudostreamer corona is produced in water, comparatively poor shock waves are created, a relative amount of bubbles observes, ultraviolet UV (poor to moderate) (excellent solution conductivity), and this system causes to formation of radicals and reactive species in the narrow area close to the corona discharge electrodes

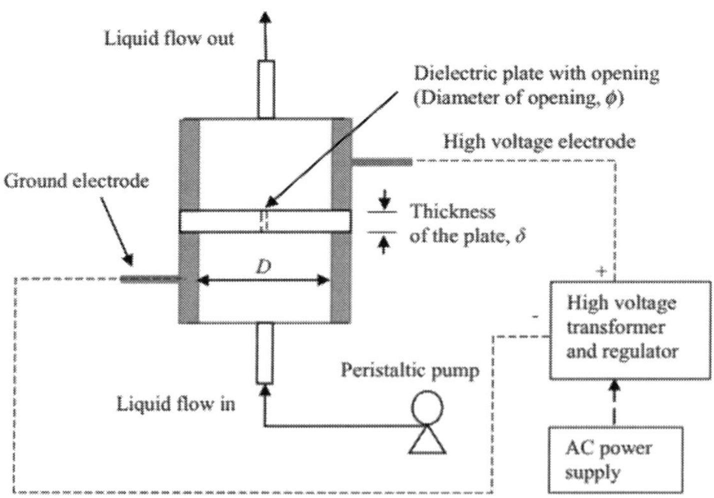

Figure 3.5 A schematic of an electrohydraulic discharge process for handling high-conductivity wastewaters (Wu & Krosuri, 2020).

(Wang, Li, Xie, & Wu, 2008). The pulsed arc system works at 10^{-2}–10^{-3} Hz, and its peak current is > 1 kA, with a voltage increase in the degree of microsecond. An arc channel causes a strong shock wave in the cavitation area and the gas in the bubble is ionized (plasma bubbles). Additional supercritical water conditions also occur. Strong UV emission and high radical concentration are found only in the short-lived cavitation area. Corona discharge is more sensitive to electrical conductivity comparing to pulsed arc. In order to apply high voltage electric discharge to solve ecological problems, it is needed to examine the kinds of chemical reactions caused by the discharge and the impacts of physical procedures (such as shock waves, cavitation, and light emission) in promoting optimal chemical reactions.

In short, the potential benefits of the electrical discharge process (such as electric discharge in water, in high water and at the same time in water and gas) are:

(1) Direct synthesis of a variety of highly active chemical species, such as molecules and radicals, elimination of the need for external resources of hydrogen peroxide production, ozone and other high reactive components;

(2) Improvement and facilitate the gas phase reaction (a) quench the products of the gas phase reaction through transferring to the liquid; and (b) species of the reactions created in the liquid phase and then transported to the gas phase;

(3) Improvement and facilitate the reaction of the liquid phase (a) Adsorption of reactive species from gas phase; and (b) Transferring volatile products resulted from liquid reaction to the gas phase;

(4) Simultaneously facilitating and increasing the reaction of the gas and liquid phase;

(5) Controlling the relative quantities of reactive species by adjusting the used electric field and the flow of gas/liquid rates;

(6) Increasing the transfer rate of gas/liquid through (a) electrohydrodynamic flow at the gas/liquid boundary; and (b) reducing the bubble dimension when the gas is introduced via a high voltage discharge electrode;

(7) Producing UV light, shock waves, and possible critical conditions located on a nonhomogeneous discharge channel; and,

(8) Catalytic impacts of pulsed electric discharge combined with different additive particles like activated carbon, zeolites, photo catalyst, transition metals and other heterogeneous catalysts, like those that are coated on the electrodes.

Designing and analyzing electrical discharge reactors of liquid reactions are a very complicated problem owing to the variety of physical and chemical parameters associated with it (Liu & Jiang, 2005). In addition, these physical and chemical procedures relate to the reactor characteristics, such as the position of the electrode, the electrode material, the existence of the gas and liquid phase, and the conductivity and composition of the solution. Currently, the knowledge of all these physical and chemical parameters is not enough to design and analyze predicted and precise reactors, but significant improvements have been made in this field.

3.6 Magnetic adsorption separation (MAS) process

Fast industrialization and technical progress are main reasons for the large volume of effluents include a mix of harmful heavy metals like lead (Pb), chromium (Cr), copper (Cu), zinc (Zn), nickel (Ni), and mercury (Hg), among others (Baziar et al., 2017). Heavy metals, unlike organic contaminants, are not bio-degradable, which is why they lead to severe ecological and public health issues. In recent years, scientists of the environmental remediation field have demonstrated an enhanced concern for the mobile and very harmful hexavalent chromium (Cr (VI)) that produced from electroplating, textile, electrolytic plants and acid plants (Zhou et al., 2018). According to severe ecological regulations like the American Environmental Protection Agency (EPA), the maximum discharged limit (MDL) of Cr (VI) are only 0.15 and 0.05 mg/L in both surface and portable water, respectively (Vengosh et al., 2016). Prior discharging into municipal sewer lines, industries must reduce the amount of chromium (VI) in their wastewater to suitable concentration levels. Removal Chrome (VI) from the aqueous environments are mostly carried out through traditional methods like discharge, ion exchange, electrocoagulation, biological procedures and adsorption. Although these conventional physical, chemical, and biological approaches sometimes suffer from rapid reaction conditions, high operating costs, and low performance of treatment, particularly by biological approaches (Dinker & Kulkarni, 2015; Pradhan, Sukla, Sawyer, & Rahman, 2017). Among the methods mentioned above, adsorption is the most used technique for chromium (VI) removal from wastewater, as it has a spectral advantage like cost-effectiveness, easy operation, technical capabilities, and does not produce additional waste. The effectiveness of any adsorption system greatly affects the physical-chemical properties of adsorbent, the improvement, and production of novel sorbents for removal of chromium ions (VI) from aqueous solutions. Nanostructured sorbents are known as potent options for reducing sewer pollutants due to their large surface area, short diffusion pathways and excellent adsorption capacity. Although recovering the adsorbate-loaded nanoscale sorbents is difficult, particularly in the flow-through system applications. Available conventional methods of separation, like centrifugation, sedimentation, filtration, and fixed beds with operating costs and maintenance, and reducing the diameter of the pressure, can eliminate the equipment (Pakade, Tavengwa, & Madikizela, 2019). Besides reducing the internal particle diffusion, optimum particle design and geometry of bed, inherent limitations have been introduced to further improve the use of the fixed classic bed filter configuration. Applying nanomagnetic sorbents, particularly those comprising iron oxide nanoparticles in solving environmental issues, is one of the new techniques that has been considerably interested in recent years because of their simple separation from water using an external magnetic field after adsorption (Fig. 3.6). Though unprotected magnetic nanoparticles are oxidized by exposing to the air and tend to be simply assembled in water solution systems, then their specific water level decreases and, as a result, their adsorption capacity is reduced (Eyvazi, Jamshidi-Zanjani, & Darban, 2019). Consequently, for practical applications of nanoparticles in diverse areas, their stability is necessary with surface modification.

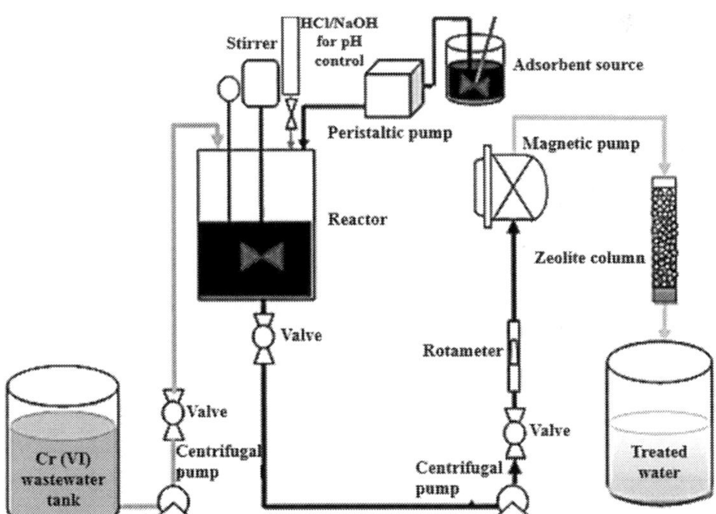

Figure 3.6 A schematic of a MAS process for extracting Cr(VI) from aqueous solution (Muliwa, Leswifi, Onyango, & Maity, 2016).

A recent method is functionalizing magnetic nanoparticles with polymers, in particular polymers like olyaniline, polythiophene and polypyrrole (Bohara, Thorat, & Pawar, 2016). Polymer conductivity is very attractive and has a variety of applications in batteries, molecular electronics, ion exchanger materials, and chemical sensors (Muñoz-Bonilla, Sánchez-Marcos, & Herrasti, 2017). Among conductive polymers, polypyrrole (PPy) and its derivatives have been remarkably interested as they can probably be used as excellent ion exchange materials for selective removal of ions from water (Chigondo, Paumo, Bhaumik, Pillay, & Maity, 2019). Furthermore, in the case of stable, well-tailored, polymer-based nanocomposites, a balance must be struck between extracting pollutants from aqueous solutions and separating adsorbent materials from the wet environment for reuse and thus avoiding their spread to the environment. Insulating magnetic nanocomposite particles from aqueous solutions using magnetic separation systems, which is simple, fast, and accurate compared with other separation systems compared with other solid-liquid separation equipment. However, to extract realistic data that is industrially adjusted, magnetic separation instruments should work continuously and particles must have effective magnetic features to ease their separation from a flowing stream. Several scientific reports indicate that magnetic adsorption and separation are mostly restricted to bench scales and batch operation modes, and therefore the data achieved does not essentially reveal a true industrial adjustment operation. Furthermore, magnetic particle capture with electromagnetic or high temperature superconducting magnets obtained, which are not cost-effective that the first one consumes electricity, and the latter needs cooling to the temperature of the cryogenic temperatures.

3.6.1 Magnetic adsorbents

Magnetic particle technology is one of several new and innovative ways in which environmental issues have been addressed in recent years (Yang, Wang, & Liu, 2020). These magnetic particles can be isolated from the solution using a simple magnetic procedure, which is done with contamination removal from the wastewater (Brião, Resende de Andrade, Gurgel Carlos da Silva, & Vieira, 2020). Additionally, suitable stripping agents can be used to remove adsorbed metals from the adsorbent surface and concentrate them in a small volume, and then they can be reused. However, the challenge related to these magnetic adsorbents is limited by their low adsorption capacity or slow adsorption because of their small areas or their porous features, and thus limits the wide application of these magnetic particles (Reddy & Yun, 2016). Magnetic nanoparticles have attracted many attentions in the past decades to counteract the defects associated with magnetic particles. In addition, the presence of large-scale magnetic nanoparticles in highly dispersed water has the super aeromagnetic properties, which can be simply separated from liquid solutions by means of an external magnetic field.

3.6.2 Magnetic nano-adsorbent in sewer treatment application

Hu, Lo and Chen (2005) reported the using surface-modified Jacobite ($MnFe_2O_4$) as a strong sorbent for removing and recycling chromium (VI) from synthetic wastewater (Hu et al., 2005). They synthesized 10 nm of functionalized $MnFe_2O_4$ nanoparticles using the sedimentary method and Redox surface reaction. The kinetic study showed that at pH of 2, the equilibrium time for chromium (VI) adsorption to modified $MnFe_2O_4$ nanoparticles was 5 min, and the adsorption data were well adapted to the Langmuir model. Also, the maximum adsorption capacity was 31.5 mg chromium (VI)/g$MnFe_2O_4$, which was compared with other conventional sorbents like activated carbon and sawdust. They investigated mostly the impacts of ligands (EDTA, SO_4^{2-}, NH_4^+) and ionic strength in the pH between 2 and 10, and found that EDTA and SO_4^{2-} prevented the adsorption of chromium (VI) in the total pH, while NH_4^+ adsorbed chromium (VI) at pH higher than 6.5. $MnFe_2O_4$ nanoparticles can be reused after regeneration without losing the adsorption capacity. One of interesting observations was that the $MnFe_2O_4$ nanoparticles were simply isolated from the aqueous solution using a simple magnetic procedure following adsorption of chromium (VI). Hu, Lo, and Chen (2007) conducted a comparative study of magnetic nanoparticles such as $CoFe_2O_4$, $MnFe_2O_4$, $MgFe_2O_4$, $ZnFe_2O_4$, $CuFe_2O_4$, and $NiFe_2O_4$ to remove chromium (VI) (Hu et al., 2007). The removal of chromium (VI) as a function of contact time in optimal conditions was investigated and the results showed that the adsorption rate of chromium (VI) increases with increasing contact time and slowly decreases in less than 60 min. The decreasing order in terms of chromium (VI) removal efficiency by these adsorbents is as follows:

$MnFe_2O_4 > MgFe_2O_4 > nFe_2O_4 > CuFe_2O_4 > NiFe_2O_4 > CoFe_2O_4$
The equilibrium time for chromium (VI) adsorption is 5, 45, 30, 20, 15, and 60 min, respectively.

3.6.3 Physics and fluid dynamics of magnetic separation

For magnetic separation, different force balances between competing forces on a magnetic particle must be considered. These forces include hydrodynamics drag due to the flow velocity, magnetic force caused by the applied field, centrifugal force, gravitational force, diffusion force and interparticle forces such as dipole-dipole interaction, interactions between two Helmholtz layers and van Der Waals attraction (Ge, Encinas, Araujo, & Song, 2017). By approaching the size of the particle to the nanometer, ductile forces play an important role because the required energy to move a particle is comparable with the thermal energy of Brownian motion, so a number of density gradients are created (Hui et al., 2018). Since the magnetic and dipole-dipole interactions and van Der Waals attraction forces helped to the separation processes; while the amounts of the magnetic (F_m), drag (F_h), diffusional (F_d), gravitational (F_g), and centrifugal (F_c) forces are considered, they are considered for force balance. If magnetic particles are to be successfully collected, attraction magnetic force must dominate all competing forces, because the particle suspension flows through the separator.

3.7 High gradient magnetic separation (HGMS)

HGMS (Fig. 3.7) is a powerful and stable technique used to remove ferromagnetic and ultraparamagnetic solids from low viscosity aqueous solutions. This technology is widely used in the mineral industry for removal of metals from industrial wastewater, as well as in the bio separation and then in sewer treatment (Ge et al., 2017). The separation of particles from the fluid carried out by the induced magnetic force due to the distortion of external magnetic fields by ferromagnetic matrices. HGMS provides the possibility of separating magnetic and nonmagnetic particles in the submicron and micron-scale because these particles tend to agglomerate (Mirshahghassemi, Ebner, Cai, & Jamie, 2017). The main advantages of this technique are presented in comparison with the filtration technology of the submerged bed as the conventional separation method, although both are very similar to each other. Separation using HGMS has high pore volume of the filter matrix than common deep bed filtration systems, which results in almost no significant pressure drop. In addition, lower operating costs and better life time are also realized, as cleansing the separator using back-flushing is possible.

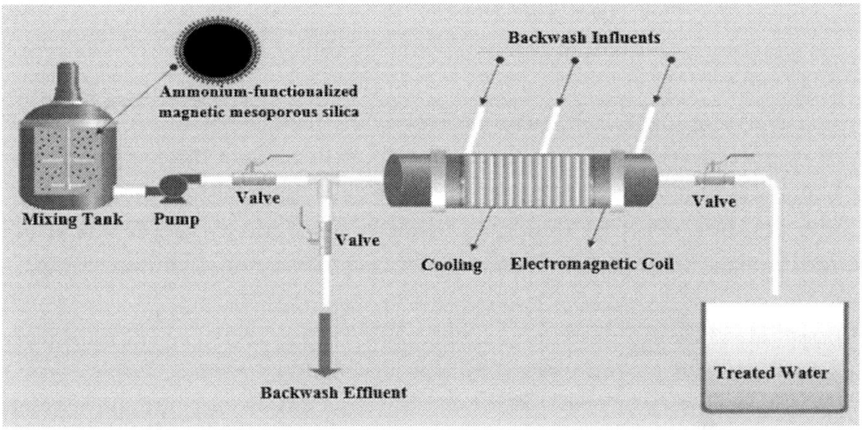

Figure 3.7 A schematic of an HGMS process for removal of nitrate from aqueous solution (Kheshti, Ghajar, Ali & Kheshti, 2019).

3.8 Magnetic induction swing adsorption (MISA)

Over the recent decades, scientists have reported a rise in the amount of anthropogenic carbon dioxide (CO_2) in the atmosphere (Lin et al., 2020). In 2012, more than 520 quadrillion Btu energy was consumed in all over the world that caused the release of more than 32 billion tons of carbon dioxide in the atmosphere as a result of using remarkable percentage of fossil based fuels (Zhu, Tian, & Hao, 2020). Postcombustion carbon capture and storage (CCS) is known as a potential procedure that can help to global greenhouse gas emission reductions purposes (Sadiq et al., 2016). Using the porous sorbent material or the chemical adsorption procedure for CCS from large point resources is still to gain a wide range of application because of the energy losses related with the implementation of these procedures. Recovery of sorbents through TSA and reformation of sorbents through PSA adsorption are two energy-consuming procedures that are associated with an increase about 40%–25% in the energy demand of a plant. The CO_2 chemical adsorption with amine-based solvents is another effective for CCS processes, but the reformation procedure needs large quantities of energy to release adsorbed CO_2 with corrosive steam and volatile amine, which has serious ecological concerns (Sadiq, Suzuki, & Hill, 2018). The metal-organic frameworks (MOFs) create crystalline nanoparticles with exceptional porosity, which are used for storing energy-related gases

(H_2 and CH_4), separation and analysis, therefore, the adjustable pore surfaces, small density, thermal and mechanical stability, MOFs can be designed for replacement of zeolite and active carbon in gas storage and separation programs (Li et al., 2017). Numerous MOFs have been introduced with several appropriate features for separating CO_2, storage of methane and storage of hydrogen. The capability of the MOF regeneration through mild trigger conditions makes them more suitable and attractive for separating and storing gas (Sadiq et al., 2020). The incorporation of magnetic nanoparticles in MOF adsorbents enables remote and rapid heat generation and transfer in MISA as compared to the gradual heat transfer process during regeneration in TSA processes (Fig. 3.8).

A novel method that includes MOFs with a variety of applied materials designed to expand and/or enhance the features of MOFs is presently being studied by several groups. A group of such magnetic materials, nanoscale or microparticles with MOF crystals are magnetic frame composites (MFCs) (Han, Ge, Sun, Gao, & Zheng, 2019). The combination of MOFs with magnetic nanoparticles which can produce heat in the face of an alternating current (AC) magnetic field can remarkably decrease energy consumption related with sorbent reformation in CCS procedures.

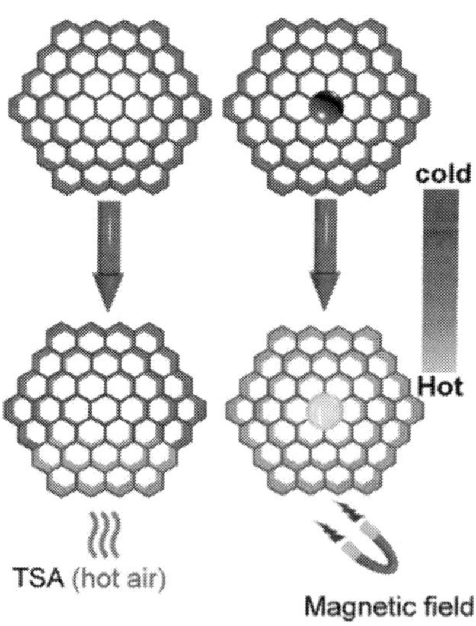

Figure 3.8 Comparison of heat transfer mode between TSA and MISA in adsorbents (Sadiq et al., 2020).

3.9 Conclusion

The current improvements of electrical-assisted and magnetic-assisted sorption processes for the removal of various contaminants from gas and liquid phases were presented. The role of efficient sorption processes such as thermal swing adsorption, electrothermal swing adsorption, pressure swing adsorption, electrohydraulic discharge, magnetic adsorption, high gradient magnetic separation and magnetic induction swing adsorption, on the sorption performance, were evaluated and well discussed.

References

Abdullah, A., Idris, I., Shamsudin, I. K., & Othman, M. R. (2019). Methane enrichment from high carbon dioxide content natural gas by pressure swing adsorption. *Journal of Natural Gas Science and Engineering, 69*, 102929.

Ajenifuja, A., Joss, L., & Jobson, M. (2020). A new equilibrium shortcut temperature swing adsorption model for fast adsorbent screening. *Industrial and Engineering Chemistry Research, 59*, 3485–3497.

Aleghafouri, A., & Davoudi, M. (2018). Modeling and simulation of a pressure–temperature swing adsorption process for dehydration of natural gas. *Adsorption, 24*, 121–133.

Ambrożek, B., & Zwarycz-Makles, K. (2014). Theoretical and experimental studies of the recovery of volatile organic compounds from waste air streams in the thermal swing adsorption system with closed-loop regeneration of adsorbent. *Energy Conversion and Management, 85*, 646–654.

Banerjee, D., Simon, C. M., Elsaidi, S. K., Haranczyk, M., & Thallapally, P. K. (2018). Xenon gas separation and storage using metal-organic frameworks. *Inside Cosmetics, 4*, 466–494.

Baziar, M., Ali, A., Karimaei, M., Gupta, V. K., Agarwal, S., Sharafi, K., et al. (2017). MWCNT-Fe_3O_4 as a superior adsorbent for microcystins LR removal: Investigation on the magnetic adsorption separation, artificial neural network modeling, and genetic algorithm optimization. *Journal of Molecular Liquids, 241*, 102–113.

Benson, T. J., & George, C. E. (2005). Cellulose based adsorbent materials for the dehydration of ethanol using thermal swing adsorption. *Adsorption, 11*, 697–701.

Bohara, R. A., Thorat, N. D., & Pawar, S. H. (2016). Role of functionalization: Strategies to explore potential nano-bio applications of magnetic nanoparticles. *RSC Advances, 6*, 43989–44012.

Bonnissel, M., Luo, L., & Tondeur, D. (2001). Compacted exfoliated natural graphite as heat conduction medium. *Carbon, 39*, 2151–2161.

Brião, G.de V., Resende de Andrade, J., Gurgel Carlos da Silva, M., & Vieira, M. G. A. (2020). Removal of toxic metals from water using chitosan-based magnetic adsorbents. A review. *Environmental Chemistry Letters*, 1–24.

Chigondo, M., Paumo, H. K., Bhaumik, M., Pillay, K., & Maity, A. (2019). Magnetic arginine-functionalized polypyrrole with improved and selective chromium (VI) ions removal from water. *Journal of Molecular Liquids, 275*, 778–791.

Delgado, J. A., Águeda, V. I., Uguina, M. A., Sotelo, J. L., Brea, P., & Grande, C. A. (2014). Adsorption and diffusion of H_2, CO, CH_4, and CO_2 in BPL activated carbon and 13X zeolite: Evaluation of performance in pressure swing adsorption hydrogen purification by simulation. *Industrial and Engineering Chemistry Research, 53*, 15414–15426.

Dinker, M. K., & Kulkarni, P. S. (2015). Recent advances in silica-based materials for the removal of hexavalent chromium: A review. *Journal of Chemical and Engineering Data, 60*, 2521–2540.

Emamipour, H., Hashisho, Z., Cevallos, D., Rood, M. J., Thurston, D. L., Hay, K. J., et al. (2007). Steady-state and dynamic desorption of organic vapor from activated carbon with electrothermal swing adsorption. *Environmental Science and Technology, 41*, 5063–5069.

Eyvazi, B., Jamshidi-Zanjani, A., & Darban, A. K. (2019). Immobilization of hexavalent chromium in contaminated soil using nano-magnetic $MnFe_2O_4$. *Journal of Hazardous Materials, 365*, 813–819.

Ge, W., Encinas, A., Araujo, E., & Song, S. (2017). Magnetic matrices used in high gradient magnetic separation (HGMS): A review. *Results in Physics*, 7, 4278–4286.

Ge, K., Wu, Y., Wang, T., & Wu, J. (2019). Humidity swing adsorption of H_2S by fibrous polymeric ionic liquids (PILs). *Separation and Purification Technology, 217*, 1–7.

Hanchate, N., Kulshreshtha, P., & Mathpati, C. S. (2019). Optimization, scale-up and cost estimation of dehydration of ethanol using temperature swing adsorption. *Journal of Environmental Chemical Engineering*, 7, 102938.

Han, L.-J., Ge, F.-Y., Sun, G.-H., Gao, X.-J., & Zheng, H.-G. (2019). Effective adsorption of Congo red by a MOF-based magnetic material. *Dalton Transactions, 48*, 4650–4656.

Hui, M., Pu, S., Hou, Y., Zhu, R., Anatoly, Z., & Chu, W. (2018). A highly efficient magnetic chitosan "fluid" adsorbent with a high capacity and fast adsorption kinetics for dyeing wastewater purification. *Chemical Engineering Journal, 345*, 556–565.

Hu, J., Lo, I. M. C., & Chen, G. (2005). Fast removal and recovery of Cr (VI) using surface-modified jacobsite ($MnFe_2O_4$) nanoparticles. *Langmuir, 21*, 11173–11179.

Hu, J., Lo, I. M. C., & Chen, G. (2007). Comparative study of various magnetic nanoparticles for Cr (VI) removal. *Separation and Purification Technology, 56*, 249–256.

Jiang, Y., Ling, J., Xiao, P., He, Y., Zhao, Q., Zheng, C., et al. (2018b). Simultaneous biogas purification and CO_2 capture by vacuum swing adsorption using zeolite NaUSY. *Chemical Engineering Journal, 334*, 2593–2602.

Jiang, L., Roskilly, A. P., & Wang, R. Z. (2018). Performance exploration of temperature swing adsorption technology for carbon dioxide capture. *Energy Conversion and Management, 165*, 396–404.

Jiang, L., Wang, R. Q., Gonzalez-Diaz, A., Smallbone, A., Lamidi, R. O., & Roskilly, A. P. (2020). Comparative analysis on temperature swing adsorption cycle for carbon capture by using internal heat/mass recovery. *Applied Thermal Engineering, 169*, 114973.

Jiang, B., Zheng, J., Qiu, S., Wu, M., Zhang, Q., Yan, Z., et al. (2014). Review on electrical discharge plasma technology for wastewater remediation. *Chemical Engineering Journal, 236*, 348–368.

Johnsen, D. L., Emamipour, H., Guest, J. S., & Rood, M. J. (2016). Environmental and economic assessment of electrothermal swing adsorption of air emissions from sheet-foam production compared to conventional abatement techniques. *Environmental Science and Technology, 50*, 1465–1472.

Joss, L., Gazzani, M., & Mazzotti, M. (2017). Rational design of temperature swing adsorption cycles for post-combustion CO_2 capture. *Chemical Engineering Science, 158*, 381–394.

Kannan, V. S., Arjunan, T. V., & Vijayan, S. (2020). Experimental investigation of temperature swing adsorption system for air dehumidification. *Heat and Mass Transfer*, 1–13.

Kheshti, Z., Ghajar, K. A., Ali, A., & Kheshti, M. R. (2019). High-Gradient Magnetic Separator (HGMS) combined with adsorption for nitrate removal from aqueous solution. *Separation and Purification Technology, 212*, 650–659.

Lei, M., Vallières, C., Grévillot, G., & Latifi, M. A. (2013). Thermal swing adsorption process for carbon dioxide capture and recovery: Modeling, simulation, parameters estimability, and identification. *Industrial and Engineering Chemistry Research, 52*, 7526–7533.

Lin, X., Shao, B., Zhu, J., Pan, F., Hu, J., Wang, M., et al. (2020). In situ electromagnetic induction heating for CO_2 temperature swing adsorption on magnetic Fe_3O_4/N-doped porous carbon. *Energy and Fuels, 34*, 14439–14446.
Li, H., Sadiq, M. M., Suzuki, K., Falcaro, P., Hill, A. J., & Hill, M. R. (2017). Magnetic induction framework synthesis: A general route to the controlled growth of metal–organic frameworks. *Chemistry of Materials, 29*, 6186–6190.
Liu, F., Chen, S., & Gao, Y. (2017). Synthesis of porous polymer based solid amine adsorbent: Effect of pore size and amine loading on CO_2 adsorption. *Journal of Colloid and Interface Science, 506*, 236–244.
Liu, Y. J., & Jiang, X. Z. (2005). Phenol degradation by a nonpulsed diaphragm glow discharge in an aqueous solution. *Environmental Science and Technology, 39*, 8512–8517.
Li, H., Wu, C., Hao, Z., Li, X., & Gao, X. (2019). Process intensification in vapor–liquid mass transfer: The state-of-the-art. *Chinese Journal of Chemical Engineering, 27*, 1236–1246.
Locke, B. R., Sato, M., Sunka, P., Hoffmann, M. R., & Chang, J.-S. (2006). Electrohydraulic discharge and nonthermal plasma for water treatment. *Industrial and Engineering Chemistry Research, 45*, 882–905.
Luo, L. (2013). Intensification of adsorption process in porous media. In *Heat and mass transfer intensification and shape optimization*. Springer.
Mallouk, K. E., Johnsen, D. L., & Rood, M. J. (2010). Capture and recovery of isobutane by electrothermal swing adsorption with post-desorption liquefaction. *Environmental Science and Technology, 44*, 7070–7075.
Mirshahghassemi, S., Ebner, A. D., Cai, B., & Jamie, R. L. (2017). Application of high gradient magnetic separation for oil remediation using polymer-coated magnetic nanoparticles. *Separation and Purification Technology, 179*, 328–334.
Mouele, E. S. M., Tijani, J. O., Fatoba, O. O., & Petrik, L. F. (2015). Degradation of organic pollutants and microorganisms from wastewater using different dielectric barrier discharge configurations—a critical review. *Environmental Science and Pollution Research, 22*, 18345–18362.
Muliwa, A. M., Leswifi, T. Y., Onyango, M. S., & Maity, A. (2016). Magnetic adsorption separation (MAS) process: An alternative method of extracting Cr (VI) from aqueous solution using polypyrrole coated Fe_3O_4 nanocomposites. *Separation and Purification Technology, 158*, 250–258.
Muñoz-Bonilla, A., Sánchez-Marcos, J., & Herrasti, P. (2017). Magnetic nanoparticles-based conducting polymer nanocomposites. In *Conducting polymer hybrids*. Springer.
Otero, M., Zabkova, M., Grande, C. A., & Rodrigues, A. E. (2005). Fixed-bed adsorption of salicylic acid onto polymeric adsorbents and activated charcoal. *Industrial and Engineering Chemistry Research, 44*, 927–936.
Pahinkar, D. G., Garimella, S., & Robbins, T. R. (2017). Feasibility of temperature swing adsorption in adsorbent-coated microchannels for natural gas purification. *Industrial and Engineering Chemistry Research, 56*, 5403–5416.
Pakade, V. E., Tavengwa, N. T., & Madikizela, L. M. (2019). Recent advances in hexavalent chromium removal from aqueous solutions by adsorptive methods. *RSC Advances, 9*, 26142–26164.
Pradhan, D., Sukla, L. B., Sawyer, M., & Rahman, P. K. (2017). Recent bioreduction of hexavalent chromium in wastewater treatment: A review. *Journal of Industrial and Engineering Chemistry, 55*, 1–20.
Qasem, N. A. A., & Ben-Mansour, R. (2018). Energy and productivity efficient vacuum pressure swing adsorption process to separate CO_2 from CO_2/N_2 mixture using Mg-MOF-74: A CFD simulation. *Applied Energy, 209*, 190–202.

Qu, Z. G., Yin, Y., Wang, H., & Zhang, J. F. (2020). Pore-scale investigation on coupled diffusion mechanisms of free and adsorbed gases in nanoporous organic matter. *Fuel, 260*, 116423.
Raganati, F., Chirone, R., & Ammendola, P. (2020). CO_2 capture by temperature swing adsorption: Working capacity as affected by temperature and CO_2 partial pressure. *Industrial and Engineering Chemistry Research, 59*, 3593–3605.
Reddy, D. H. K., & Yun, Y.-S. (2016). Spinel ferrite magnetic adsorbents: Alternative future materials for water purification? *Coordination Chemistry Reviews, 315*, 90–111.
Ren, F., Ge, L., Arami-Niya, A., Rufford, T. E., Xing, H., & Rudolph, V. (2019). Gas storage potential and electrohydraulic discharge (EHD) stimulation of coal seam interburden from the Surat Basin. *International Journal of Coal Geology, 208*, 24–36.
Ried, T., Salazar Duarte, G., & Hinrichsen, O. (2020). Experimental validation of a multidimensional model for an indirect temperature swing adsorption unit. *Chemie Ingenieur Technik, 92*, 711–719.
Sadiq, M. M., Li, H., Hill, A. J., Falcaro, P., Hill, M. R., & Suzuki, K. (2016). Magnetic induction swing adsorption: An energy efficient route to porous adsorbent regeneration. *Chemistry of Materials, 28*, 6219–6226.
Sadiq, M. M., Konstas, K., Falcaro, P., Hill, A. J., Suzuki, K., et al. (2020). Engineered porous nanocomposites that deliver remarkably low carbon capture energy costs. *Cell Reports Physical Science, 1*, 100070.
Sadiq, M. M., Suzuki, K., & Hill, M. R. (2018). Towards energy efficient separations with metal organic frameworks. *Chemical Communications, 54*, 2825–2837.
Schell, J., Casas, N., Marx, D., & Mazzotti, M. (2013). Precombustion CO_2 capture by pressure swing adsorption (PSA): Comparison of laboratory PSA experiments and simulations. *Industrial and Engineering Chemistry Research, 52*, 8311–8322.
Siqueira, R. M., Freitas, G. R., Peixoto, H. R., Nascimento, J. F do, Musse, A. P. S., Torres, A. E., et al. (2017). Carbon dioxide capture by pressure swing adsorption. *Energy Procedia, 114*, 2182–2192.
Tang, Y., Hou, C., He, Y., Wang, Y., Chen, Y., & Rui, Z. (2020). *Review on pore structure characterization and microscopic flow mechanism of CO_2 flooding in porous media*. Energy Technology.
Tao, L., Xiao, P., Qader, A., & Webley, P. A. (2019). CO_2 capture from high concentration CO_2 natural gas by pressure swing adsorption at the CO2CRC Otway site, Australia. *International Journal of Greenhouse Gas Control, 83*, 1–10.
Thouchprasitchai, N., Pintuyothin, N., & Pongstabodee, S. (2018). Optimization of CO_2 adsorption capacity and cyclical adsorption/desorption on tetraethylenepentamine-supported surface-modified hydrotalcite. *Journal of Environmental Sciences, 65*, 293–305.
Vengosh, A., Coyte, R., Karr, J., Harkness, J. S., Kondash, A. J., Ruhl, L. S., et al. (2016). Origin of hexavalent chromium in drinking water wells from the piedmont aquifers of North Carolina. *Environmental Science and Technology Letters, 3*, 409–414.
Wang, H., Li, J., Xie, Q., & Wu, Y. (2008). Enhanced generation of oxidative species and phenol degradation in a discharge plasma system coupled with TiO_2 photocatalysis. *Applied Catalysis B: Environmental, 83*, 72–77.
Wiheeb, A. D., Helwani, Z., Kim, J., & Othman, M. R. (2016). Pressure swing adsorption technologies for carbon dioxide capture. *Separation and Purification Reviews, 45*, 108–121.
Wu, S., & Krosuri, A. (2020). A novel continuous-flow electrohydraulic discharge process for handling high-conductivity wastewaters. *International Journal of Environmental Science and Technology, 17*, 615–624.
Wurzbacher, J. A., Gebald, C., Piatkowski, N., & Steinfeld, A. (2012). Concurrent separation of CO_2 and H_2O from air by a temperature-vacuum swing adsorption/desorption cycle. *Environmental Science and Technology, 46*, 9191–9198.

Yang, W., Wang, Z., & Liu, Y. (2020). Review on magnetic adsorbents for removal of elemental mercury from flue gas. *Energy and Fuels, 34*, 13473–13490.
Zhao, R., Deng, S., Wang, S., Zhao, L., Zhang, Y., Liu, B., et al. (2018). Thermodynamic research of adsorbent materials on energy efficiency of vacuum-pressure swing adsorption cycle for CO_2 capture. *Applied Thermal Engineering, 128*, 818–829.
Zhao, R., Liu, L., Zhao, L., Deng, S., & Li, H. (2018). Thermodynamic analysis on carbon dioxide capture by Electric Swing Adsorption (ESA) technology. *Journal of CO_2 Utilization, 26*, 388–396.
Zhao, R., Liu, L., Zhao, L., Deng, S., Li, S., & Zhang, Y. (2019). A comprehensive performance evaluation of temperature swing adsorption for post-combustion carbon dioxide capture. *Renewable and Sustainable Energy Reviews, 114*, 109285.
Zhou, L., Zhang, G., Wang, M., Wang, D., Cai, D., & Wu, Z. (2018). Efficient removal of hexavalent chromium from water and soil using magnetic ceramsite coated by functionalized nano carbon spheres. *Chemical Engineering Journal, 334*, 400–409.
Zhu, C., Tian, H., & Hao, J. (2020). Global anthropogenic atmospheric emission inventory of twelve typical hazardous trace elements, 1995–2012. *Atmospheric Environment, 220*, 117061.

SECTION TWO

Passive mechanisms

CHAPTER FOUR

Sorption processes using nanostructures and nanofluids

4.1 Introduction

Nowadays, the nanostructured materials have emerged significantly for various applications in many fields during the manufacturing, processing, and packaging of products (Sun, Zhang, Cui, Yang, & Liang, 2019). These materials have a size with one dimension at least less than 100 nm, while the have been categorized in various types, such as nanoparticle, nanotube, nanowire, and nanofiber (Chen, Fan, et al., 2018; Ha, Su, Xing, Zhang, & Xiong, 2017). The nanostructures have the higher performance for different processes as compared to other bulk materials. In fact, they are characterized by different physical, chemical, and biological properties than their normal size equivalents (Thangadurai, Manjubaashini, Thomas, & Maria, 2020). Metals, metal oxides, polymers and ceramics, and carbon derivatives are known as such materials that have a higher ratio of surface area to particle size at the nanoscale level. Since the surface area of particles increases with decreasing particle size, they have different optical, electrical, and magnetic properties than those exhibited by macroscopic particles (Zhu et al., 2018). The excellent properties of particles at the nanoscale level possibly corresponded to the increase in the number of surface atoms with the decreasing of particle size. Nanotechnology is a so-called platform technology because it can easily be coupled with other technologies and improves any existing scientific concept (Chen, Li, et al., 2018). The nanotechnology could help to decrease production costs by reducing energy consumption, reduction in environmental pollution, and enhanced production efficiency (Ibrahim, Hayyan, AlSaadi, Hayyan, & Ibrahim, 2016). This technology provides an adequate opportunity to facilitate different social problems of developing countries such as supplying clean water and the treatment of epidemic diseases (Santhosh et al., 2016).

One of the most promising applications of nanotechnology is improving the chemical processing performance. Among the most applicable processes, sorption is one of the most common and effective techniques owing to low

Intensification of Sorption Processes
ISBN: 978-0-12-821411-4
https://doi.org/10.1016/B978-0-12-821411-4.00002-X

cost, simple operation, strong practicality, environment-friendly, and the simple regeneration process (Fiyadh et al., 2019).

Adsorption is a mass transfer process that operates based on the accumulation of substances at the interface of two phases (Lima, 2018). Mass transfer can occur at liquid—liquid, gas—liquid, gas—solid, or liquid—solid interface through the physical forces or via weak chemical forces. The adsorption from solution on to a solid surface arises due to the driving forces of lyophobic character of the solute, relative to the particular solvent, or a high affinity of the solute for the solid (Tran, Wang, You, & Chao, 2017). The molecule which gets adhered on the solid surface is named the adsorbate, whereas the solid surface is called the adsorbent (Li et al., 2019). The efficiency of the adsorption process is influenced by different factors including the initial adsorbate concentration, temperature, nature of adsorbate and adsorbent, pH, the quantity of adsorbent, contact time, particle shape and size, and so forth (Maity et al., 2018). The performance of the adsorption process depends on solid—liquid equilibria which affects the mass transfer rate. The adsorption process can be operated at three states: batch, semibatch, and continuous (Jin et al., 2017; Hayati et al., 2018). The state of equilibrium is very important in the adsorption process, while it achieves when the concentration of the adsorbate on the solid surface and in the solution become constant (Raghav and Kumar, 2018). In this situation, the adsorption isotherm is estimated through the relationship between the amount of solid adsorbed and in solution. Adsorption isotherm is used for evaluation of interaction of adsorbate with adsorbent and is necessary for optimization of the adsorbent loading (Duan, Gu, Du, & Xian, 2016). The Langmuir, Freundlich, Temkin, and Sips are the most common isotherm models used for the adsorption process (Araújo et al., 2018).

Adsorption of the an adsorbate can be performed using nanofluids. A nanofluid is prepared by the dispersion of metallic or nonmetallic nanoparticles or nanofibers with a typical size of less than 100 nm in a liquid (Sidik, Mahmud Jamil, Japar, & Adamu, 2017). Many kinds of researches have focused on the nanofluids applications owing to the desirable physicochemical properties achieved by these substances. Nanofluids can improve the thermal diffusivity and conductivity, viscosity, mass transfer rate, and convective heat transfer coefficient in comparison with base fluids (Babu, Kumar, & Rao, 2017). Furthermore, they provide a large specific surface area which could significantly enhances the thermal efficiency, stability, and mass and heat transfer efficiency (Das, 2017). Accordingly, it is no surprise that the use of nanofluids will attract huge attention among academics and engineers for improving process performance. For instance, the acceptable improvement in heat transfer properties of nanofluid has caused that the researchers apply them in various engineering applications such as cooling of electronic apparatus, automotive, nuclear reactors, solar collectors, and mass transfer applications (Sajid & Ali, 2019; Gorji & Ranjbar, 2017).

Although the complete mechanism that clarify mass transfer enhancement by nanofluid is still unknown, the primary mechanisms that proposed for the description of convective mass transfer enhancement are Brownian motion, hydrodynamic effects, grazing effects or shuttle effect, bubble breaking effects, reductions in surface tension, and a decrease in film thickness (Pashaei & Ghaemi, 2020; Taheri, Mohebbi, Hashemipour, & Rashidi, 2016). Nanofluids provide faster CO_2 absorption kinetics due to the intensified mass transfer derived from the addition of the nanoparticles, compared to the conventional solvents (Ghasemi, Irani, & Tavasoli, 2020). Furthermore, nanofluids can be used in combination with other CO_2 capture technologies in order to the intensification of the different processes performance (Saidi, 2020).

4.2 Intensified-adsorption process using nanostructures

The adsorption process is performed either in single step or in combination of the steps such as film or external diffusion, pore diffusion, surface diffusion and adsorption on the pore surface (Khodadadi et al., 2018). The adsorption process has three main steps: (1) migration of the adsorbate to the adsorbent surface, (2) dissociation (or deprotonation) of complexed aqueous adsorbate, and (3) surface complexation (Kim, Jung, & Kang, 2014). Fig. 4.1 shows the accepted pathways of adsorption process, whereas the adsorbate is diffused on external surface of the adsorbent owing to diffusion potential corresponded to the concentration gradient of adsorbate and available external surface area on the adsorbent (Fig. 4.1). At first, the adsorbate diffuses on the external surface of the adsorbent, then diffuses on the available pores of the adsorbent. All the available exposed active sites are occupied during the adsorption process governed by either physisorption or chemisorption (Koca et al., 2018).

4.3 Application of nanostructures for intensification of different processes performance

4.3.1 Wastewater treatment using nanostructures

The nanotechnology as a fast-growing technology provides leapfrogging approaches for pollutants treatment by overcoming the shortcomings of

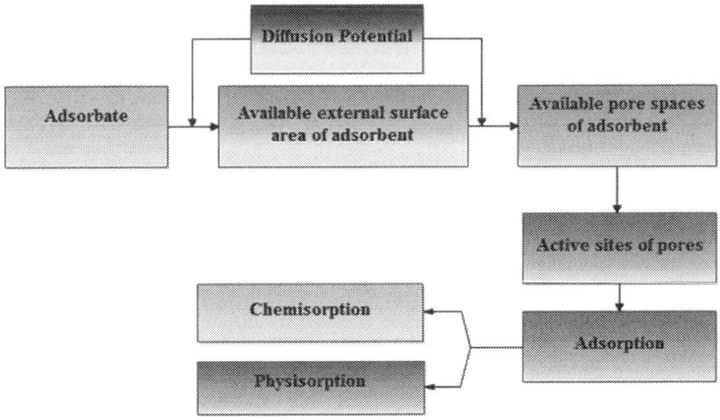

Figure 4.1 Pathways of adsorption process (Lata & Samadder, 2016).

existing processes as well as offering cost-effective procedure with superior performance (Sheikh et al., 2019). Nanomaterials can be used for remediation of aqueous environments including, water, wastewater and groundwater with five main mechanisms which are shown in Fig. 4.2.

Nanotechnology has great potential for enhances the efficiency of the wastewater treatment process and provides a sustainable approach to secure water supply (Sheikh et al., 2019). However, nanostructures suffer from some inherent technical bottlenecks such as aggregation, difficult separation, leakage into the contact water, as well as potential adverse effect imposed on ecosystem and human health. Nanostructures have used as adsorbents and photocatalyst for wastewater treatment, while they exhibited high capacity with fast kinetics toward targeted contaminants (Murgolo et al., 2019). Although there are still many obstacles to overcome for large-scale

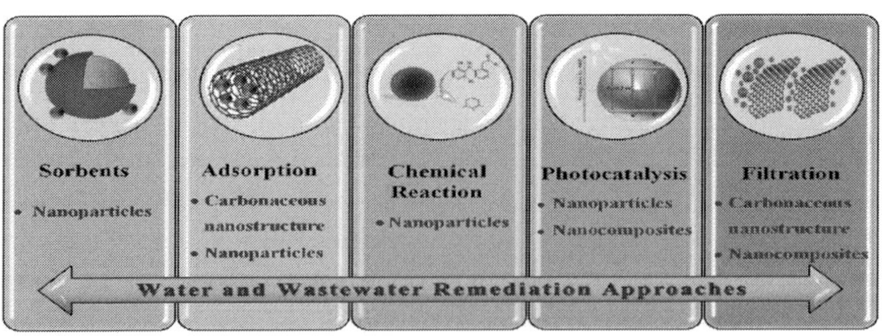

Figure 4.2 Five common approaches for remediation of water, wastewater, and groundwater remediation (Taghipour, Hosseini, & Ataie-Ashtiani, 2019).

application, nanostructures are one of the most promising candidates for the development of the sustainable next-generation water treatment technology. Nanostructures are usually unstable and tend to agglomerate owing to the van Der Waals forces and other interactions (Lau et al., 2019). Another obstacle is the recycle of nanostructures from treated water, which causes that use of these materials to be uneconomic or sometimes unfeasible. Furthermore, the transformation and toxicity of nanoparticles are not completely clarified, and it is necessary to be understood (Wan et al., 2019).

The development of nanobased composite materials which merit advantages of both the hosts and the impregnated functional nanoparticles is a great opportunity for practical applications (Navasingh et al., 2019). The dispersion and stability of the loaded nanoparticles can be facilitated using hosts like polymers, biopolymers, minerals, activated carbons, or membranes (Bharti, Thakur, Anand, & Dutta, 2020). These types of hosts improve the transfer rate or diffusion of contaminants in the hosts, which subsequently leads to intensifying the interfacial interaction. The releases of nanoparticles into the receiving water could be mitigated using nanocomposites. Nanocomposites could also effectively mitigate the release of nanoparticles into the receiving water. Furthermore, nanocomposites are a promising candidate for compatibility with existing devices like fixed-bed column or fluidized bed toward the intensification of different separation processes performance (Rezakazemi, Darabi, Soroush, & Mesbah, 2019). Fig. 4.3 shows the effects and mechanisms of various nanostructures on the adsorption of different contaminants.

The nanocomposites could be a revolution for water and wastewater treatment modes, while future research should forward the understanding of interactions between immobilized nanoparticles and hosts, sustainable techniques for host structure and nanoparticle manipulation (Gamallo, Fernández, Feijoo, & Moreira, 2020). However, the possible negative impacts on the ecosystems should be further investigated and their synthesis via green methods should also be pursued in parallel to reduce their environmental impacts (Mondal, Anweshan, & Purkait, 2020). Accordingly, practical protocols, standards, and guidelines should be assigned to the application of nanomaterials to diminish their negative impact on human health and aquatic ecosystems. Protocols and guidelines should be established to regulate the use of nanomaterials to minimize their impact to human health and aquatic environment. These protocols, standards, and guidelines help to a deep understanding of the nanostructures impact (Zhang et al., 2016). Table 4.1 represents the merits of various nanocomposites for different applications.

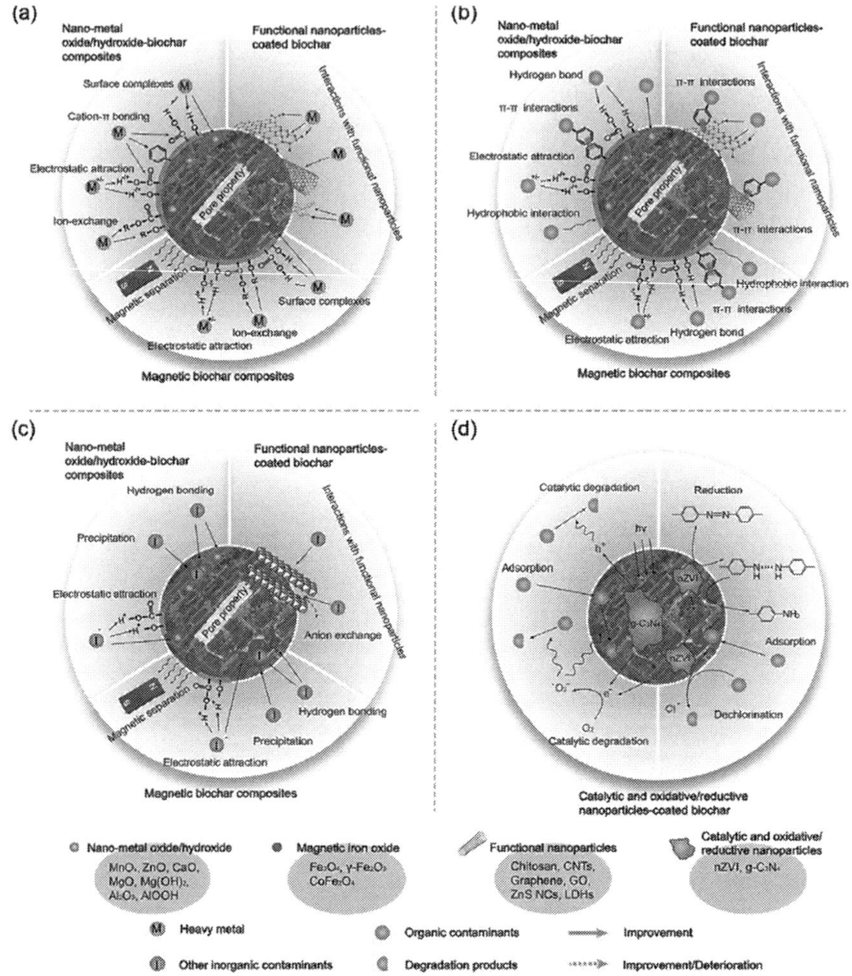

Figure 4.3 The effects and mechanisms of different nanomaterials on the removal of various contaminants: (A) adsorption of heavy metals, (B) adsorption of organic contaminants, (C) adsorption of other inorganic contaminants, and (D) simultaneous adsorption and degradation of organic contaminants (Tan et al., 2016).

As can be seen in Table 4.2. Nanometal oxide–based adsorbents can be applied for the adsorption of various organic pollutants with high performance.

4.3.2 Adsorption of heavy metals

The removal of heavy metals (metals and metalloids with an atomic density greater than 4 ± 1 g/cm^3, e.g., Cu, Zn, Hg, Cd, Pb, Sn, Fe, Mn, Ag, Cr,

Table 4.1 Different hosts for fabrication of environmental nanocomposites (Zhang et al., 2016).

Hosts	Advantages	Properties
Graphene	Laminar structures; high selectivity and transport rate toward water molecules; good mechanical strength	Adsorption, separation, antimicrobial
CNT	Nanochannel for ultrafast molecular transport; superb mechanical strength	Adsorption, sensing
Activated carbon	Low cost; high porosity; good compatibility	Adsorption
Silica	Remarkable mechanical strength; good chemical stabilities; tunable structure	Adsorption, electrode materials, (photo) catalysis
Polymer	Crosslinked structure for stable immobilization; excellent mechanical strength, chemical stability and compatibility; tunable structure; charged surface promote dispersion of nanoparticles and contaminants; ease of separation	Adsorption, sensing, (photo)catalysis, antimicrobial
Bio-polymer	Abundant and readily available resource; moderately flexible molecule; hydrophilicity, chirality, and eco-friendly	Adsorption, antimicrobial
Minerals	Low cost; chemical stability; heat resistance; eco-friendly; unique structure providing functional groups	Adsorption, in-situ remediation
Membranes	Porous structure; tunable surface properties; chemical, thermal and mechanical stability; stable nanoparticle incorporation; high compatibility	Adsorption, separation, antimicrobial, (photo)catalysis
Magnetic substrate	Ease of surface modification; easy to separate; fast kinetics	Adsorption, (photo) catalysis, coagulation

Co, Ni, As, Al, etc.) as one of the most biologically dangerous and toxic components which pollute water resources is an undeniable mission (Ni et al., 2019). The ions of such metals are known as the most widespread toxic mineral pollutants which threat soil and water systems. Different techniques including chemical precipitation, solvent extraction, membrane

Table 4.2 Removal of organic pollutants using nanometal oxide (Santhosh et al., 2016).

Adsorbents	Organic pollutants	Adsorption capacity (mg/g)	References
SiO_2/Fe_3O_4	MO	53.19	Shariati-Rad, Irandoust, Amri, Feyzi, and Ja'fari (2014)
γ-Fe_2O_3/SiO_2/chitosan composite	MO	34.29	Zhu et al. (2011)
γ -Fe_2O_3 crosslinked chitosan composite	MO	29.46	Zhua, Ru, Xiaob, and Lib (2010)
Magnetic nanopowder	Phenol	13.5	Mihoc, Ianoş, and Păcurariu (2014)
γ -Fe2O_3/2C nanocomposite	Phenol	42.34	Istratie, Stoi, Păcurariu, & Locovei (2019)
γ -Fe_2O_3/2C nanocomposite	MO	72.68	Istratie et al. (2019)
γ -Fe_2O_3	Acridine orange	59	Qadri, Ganoe, and Haik (2009)
Polyacrylic acid-Fe_3O_4	MB, crystal violet	199,116	Mak and Chen (2004), Liao, Wu, and Chen (2005)
Fe_3O_4-CTAB	Acid red 27	1.05	Zargar, Parham, and Hatamie (2009)
Chitosan-Fe_3O_4	Acid orange 12 acid green 25	1.883, 1.471	Chang and Chen (2005)
$Fe_3O_4@SiO_2$/SiCRG	Metoprolol (MTP)	447	Soares, Simões, António, Trindade, and Daniel-da-Silva (2016)
SiO_2/SiCRG	Metoprolol (MTP)	393	Soares et al. (2016)
Fe_3O_4 MNPs	Crystal violet	166.67	Muthukumaran, Sivakumar, and Thirumarimurugan (2016)

filtration, ion exchange, electrochemical removal, coagulation are used for the removal of heavy metals from water resources(Bolisetty, Peydayesh, & Mezzenga, 2019), while the advantages and disadvantages of these conventional techniques are shown in Fig. 4.4.

The conventional techniques suffer some drawbacks such as incomplete removal, high-energy requirements and availability of toxic sludge, low efficiency, sensitive operating conditions, and costly disposal (Zhu, Fan,

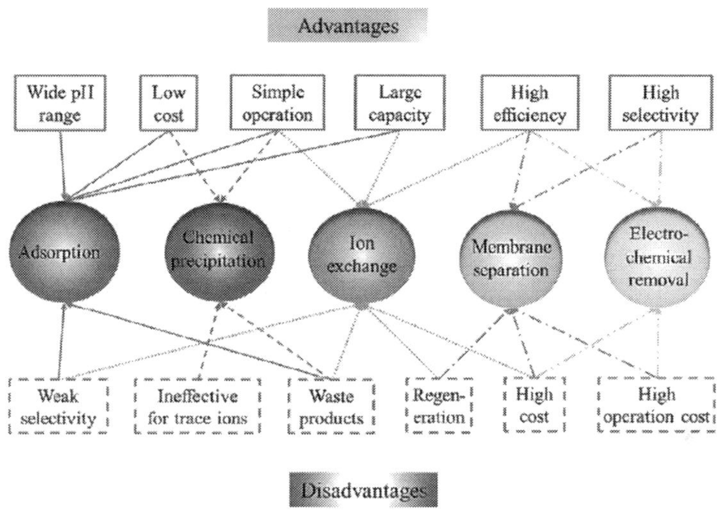

Figure 4.4 Conventional techniques for the removal of aqueous heavy metals (Xu et al., 2018).

Zhou, & Li., 2019). Among the adsorption has attracted more attention for the removal of aqueous heavy metals owing to some advantages including simple operation, low cost, good pH tolerance, and large industrial processing capacity (Yang et al., 2019). However, achieving a simple and cheap technique with high efficiency and selectivity that does not require a large investment or energy inputs yet remained as a great challenge (Jawed, Saxena, & Pandey, 2020). To overcome these drawbacks, many approaches aimed at developing cheaper and more efficient methods to improve the quality of treated effluents have been proposed. Most of them operate based on using adsorption mechanism because the adsorption has the greatest impact on transport, toxicity and biological availability of heavy metals. Furthermore, the adsorption process is easy to operate and cost-effective. The adsorption process can be along with the inverse process—desorption that facilitates the transfer of adsorbate ions from the adsorbent surface to the solution bulk(Shen et al., 2019). The reversibility of adsorption corresponds to the adsorbate amount desorbed from the adsorbent, whereas more desorbed adsorbate leads to more reversibility. The adsorption process act based on two main mechanisms: physical adsorption and chemical adsorption. The first mechanism is dependent on the substance nature, where the increase in the adsorbate concentration at the interface is due to nonspecific van Der Waals forces. The second namely chemisorption

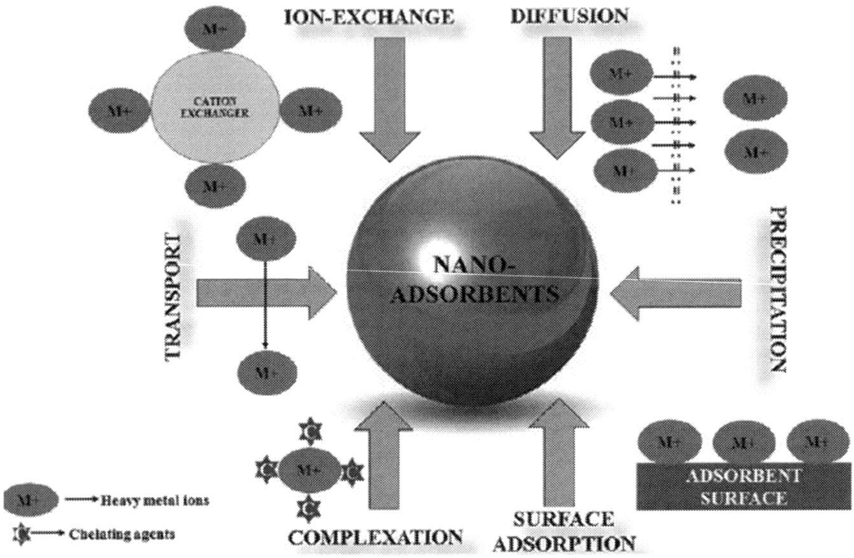

Figure 4.5 Schematic illustration of various mechanisms of adsorption of metal ions by nanoparticles (Dubey, Banerjee, Nath Upadhyay, & Sharma, 2017).

comes from chemical reactions between the adsorbate and the adsorbent which makes covalent or ionic bonds. Unlike the physical adsorption, which is weakly specific, reversible, with low thermal effect, the chemisorption is selective, usually irreversible, with large thermal effect. The mechanism involved for adsorption of sample metallic species is shown in Fig. 4.5. This mechanism has three steps as follows: film diffusion that requires transport of the adsorbate through a surface film to the exterior of the adsorbent, pore diffusion that assumes diffusion of the adsorbate within the pores of the adsorbent, and intraparticle diffusion includes adsorption of solute at the interior surfaces of the adsorbent. To have a thorough understanding of the adsorption mechanism, different analytical techniques have been used including infrared (IR) spectroscopy, X-ray diffraction (XRD), X-ray photoelectron spectroscopy (XPS), and extended X-ray absorption fine structure (EXAFS) spectroscopy. These techniques help to discover probable mechanism responsible for the adsorption which may involve either one or the combination of the two or more mechanisms such as physical adsorption, surface-complexation, ion-exchange, electrostatic interaction, and hard/soft acid-base interaction.

The adsorption is flexible in design while it is also known as an easy operation process. In this regards, adsorption represents a great performance for removal of different contaminants from wastewater effluents. One of the desirable properties of the adsorption is the reversibility, which helps the regeneration of adsorbent for reusability. The reusability is a key parameter, especially in the case of using expensive adsorbents. Therefore, it is many important to preparing novel adsorbents which have higher reusability. Nanotechnology can be used to design such adsorbents. Nanotechnology provides excellent physical, chemical, and mechanical properties to modifying conventional adsorbents or preparing alternative adsorbents, which enhances the removal efficiency (Burakov et al., 2018). Table 4.3 represents the performance of some novel adsorbent for removal of different heavy metal ions.

Table 4.3 Adsorption capacity of some adsorbents for different heavy metal ions (Burakov et al., 2018).

Adsorbent	Adsorbate	q_m
Graphene oxide (GO)/Fe_3O_4	Cu^{2+}	18.3
GO/SiO_2/Fe_3O_4	Pb^{2+} and Cd^{2+}	333.3 and 166.7
GO/Fe_3O_4/sulfanilic acid	Cd^{2+}	55.4
GO-$MnFe_2O_4$	Pb^{2+}	673
Reduced GO/$COFe_2O_4$	Pb^{2+}	299.4
GO/Mn$-$$Fe_2O_3$	Cd^{2+} and Cu^{2+}	87.2 and 129.7
GO/Fe_3O_4/sulfanilic acid	Cu^{2+}	50.7, 56.8
GO/Fe_3O_4	Cu^{2+}, Pb^{2+}, Cd^{2+}	23.1, 38.5, and 4.4
GO	Pb^{2+}	35.6
TiO_2/GO	Pb^{2+}	65.6
GO-EDTA	Pb^{2+}	479
Graphene nanosheets (GNSs)	Pb^{2+}	22.4
GNSs-500	Pb^{2+}	35.2
$SiO_2$$-$GNSs	Pb^{2+}	113.6
MnO_2/GNSs	Hg^{2+}	10.8
GO- based inverse spinel nickel ferrite composite	Cr^{6+}	45, pH 4.0
Fe(0) assembled on magnetic Fe_3O_4/graphene nanocomposites	Cr^{6+}	101, pH 3.0
Copolymer of dimethylaminoethyl methacrylate with graphene oxide	Cr^{6+}	82.4, pH 1.1
Graphene sand composite	Cr^{6+}	2859.38, pH 1.5
GO	Cr^{6+}	43.72, pH 5.0
GNSs modified with cetyltrimethylammonium bromide	Cr^{6+}	21,57, pH 2.0

4.4 Regeneration and reusability of adsorbents

One the major challenges in the adsorption processes is the recovery of the adsorbent for further reuse. The separation of the adsorbents from treated solution can be difficult and costly. Magnetic nanomaterials are a good candidate for overcoming this challenge due to their unique properties. These materials are possible to be manipulated under the impact of an external magnetic field. Recently, novel synthesis approaches have led to the preparation of superparamagnetic nanoparticles with different shell and surface modifications. Magnetic nanoparticles (MNPs) build from the magnetic elements including iron, cobalt, nickel, or their oxides like magnetite (Fe_3O_4), maghemite (γ-Fe_2O_3), nickel ferrite ($NiFe_2O_4$), and cobalt ferrite ($CoFe_2O_4$). MNPs can be coated with organic layers such as surfactants or polymers such as dextran and polyethylene glycol. Furthermore, they can be coated with inorganic components, such as metallic elements, metal oxides, activated carbon, and silica. The coating not only enables them to be stable against oxidation, corrosion, and spontaneous aggregation but also promote their physicochemical stability and to create a functionalizable surface (Gómez-Pastora, Bringas, & Ortiz, 2014). MNPs have great potential in solving environmental issues due to some of their inherent properties, as follows:

- Higher economical and environmental efficiency owing to unique physical and chemical properties such as high surface area, ease of functionalization and chemical stability.
- The superparamagnetic behavior that facilitates the design of separation and recovery processes.

The MNPs-based adsorption processes include following steps: (1) adsorption of contaminants using adsorbents, (2) magnetic recovery of the adsorbent from treated solution, (3) adsorbent regeneration for further usage, and (4) management of both the spent regeneration solution and the saturated adsorbent. It seems that the regeneration process and adsorbent reusability have received much less attention in spite of their great importance for the adsorption process economy. Accordingly, these key parameters should be considered for a design step along with the technical feasibility of MNPs for treatment processes. It is worth noting that design of efficient magnetic separation devices is a crucial stage for a sustainable process. Furthermore, the management of both the solid and liquid wastes effluent from the adsorption process is an environmental problem, which should be solved.

4.5 Sorption processes using nanofluids

The behavior of the fluids under shear stresses is investigated by the science of rheology. Fluids are categorized to Newtonian and non-Newtonian fluids, which the first are those that accept the Newton's linear friction law, while the latter does not follow such law. The non-Newtonian fluids are usually very viscous, whereas they are divided into several subgroups including Bingham's plastic, pseudo-plastic and dilatant fluids in terms of rheological behavior (Fig. 4.6). Generally, non-Newtonian fluids are examined in three groups consist of time dependent, time-independent, and viscoelastic fluids (Khodadadi et al., 2018).

The addition of materials into fluids affect the rheological behavior of the fluids. Accordingly, the nano-additives are used to prepare nanofluid for different applications. Nanoadditive refers to a particle of solid matter that has nano scales, at least in one dimension (less than 100 nm). The cylindrical nanostructures, spherical nanoparticles, and polyhedral nanoparticles are different types of nanostructures that can be prepared through various synthesis methods. As an example, the rheological behavior of nanofluids containing copper oxide and carbon nanotubes is presented in Table 4.4. Since different parameters such as the temperature, nanoparticle volume fraction, type of nanoparticle and base fluid, manufacturing method and stabilization can affect the rheological behavior of each nanofluid, small change in the experiment conditions can affect the results (Koca et al., 2018).

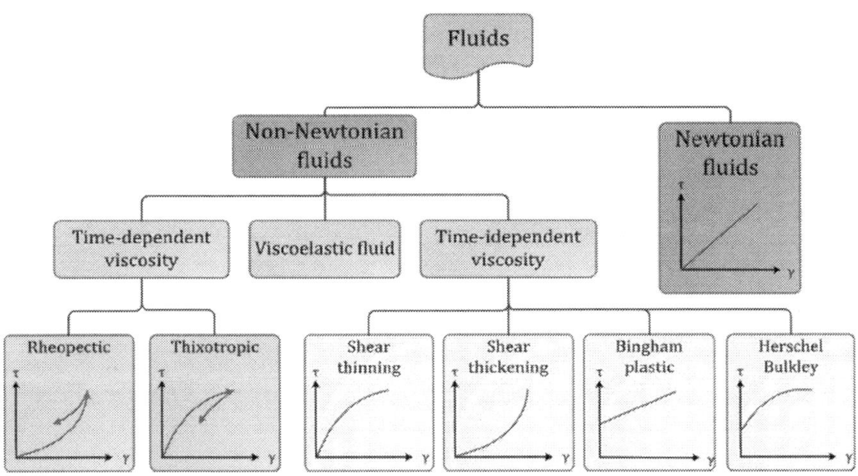

Figure 4.6 Types of fluid behavior (Khodadadi et al., 2018).

Table 4.4 A summary of studies on rheological behavior of some nanofluids (Khodadadi et al., 2018).

Nanostructure used	Base fluid	Rheology	Key results/remarks
CuO	PG:Water (60:40)	Newtonian	Relative viscosity = 1.043 at 1.2 vol.%
CuO	EG:water	Newtonian (up to 0.1%) then shear-thinning	Presence of aggregates, low shear viscosity increases with volume fraction
CuO	EG:Water (60:40)	Newtonian	Higher relative viscosity of 4.5 at −30°C and 6.12 vol.%
CuO	EG:water	Newtonian	Viscosity decreases with NP loading, relative viscosity = 0.87 at 1 vol.%
MWCNTs	Oil	Newtonian and shear-thinning	At $\varphi = 0.34$ vol.%: $\tau = 10$ pa, $\mu_{eff} = 1570$ At $\varphi = 0.34$ vol.%: $\tau = 50$ pa, $\mu_{eff} = 30.20$
MWCNTs	Water	Non-Newtonian shear-thinning	A number of mechanisms related to boundary layer thickness, microconvective effect, particle rearrangement, effects due to viscosity
MWCNTs	Silicon oil	Newtonian	At $\varphi = 0.54$ wt.%: $\mu_{eff} < 1$ (decreasing)
MWCNTs	Water	Non-Newtonian	At $\varphi = 0.5$ wt.%: $\mu_{eff} < 0.8$ (decreasing)
MWCNTs	Water	Newtonian and shear-thinning	At $\varphi = 0.55$ vol.%: $\mu_{eff} = 5.2$ with SDBS At $\varphi = 0.55$ vol.%: $\mu_{eff} = 30.20$ with Lignin

Table 4.4 A summary of studies on rheological behavior of some nanofluids (Khodadadi et al., 2018).—cont'd

Nanostructure used	Base fluid	Rheology	Key results/remarks
MWCNTs	Water	Newtonian and shear-thinning	Nanofluid viscosity increased up to 7 min of sonication and then decreased with more rise in sonication time
MWCNTs	Water	Shear-thinning	Nanofluid viscosity augmented with TX100 and CNT loading

4.5.1 Nanofluid preparation methods

Nanofluids are prepared by dispersion of nanoparticles into liquids (as the base fluids). The agglomeration is the main problem during the synthesis process. A suitable nanofluid has unique characterizations including stable suspension, low agglomeration of particles, and no chemical change of the fluid (Devendiran & Amirtham, 2016).

Various techniques are used for stabilizing the suspensions such as changing the pH value of suspension, using surface activators and/or dispersants, and using ultrasonic vibration. These techniques can affect the surface properties of the suspended particles, and suppress the formation of particle clusters to achieve stable suspensions. The choice of each technique depends on the nature of the nanoadditive as well as the base fluid properties, while the selection of suitable activators and dispersants is related to the properties of the solutions and particles. However, the addition of dispersant/surfactant changes the thermophysical properties of base fluids, while the stability of dispersed nanostructures in base fluid is evaluated by zeta potential value. High zeta potential value means good stability.

The recommended method for production of hybrid nanofluid is two-step method. Hybrid nanofluid is a blend of two or more nanostructures synthesis and dispersed in a base fluid which enhances heat and mass transfer owing to its synergistic effect compared to nanofluid containing one nanostructures. Hybrid nanofluids provide better heat and mass transfer performance than conventional fluids and nanofluids with single nanostructures. Hybrid nanofluid can replace single nanofluid due to intensification of the different processes performance (Sidik et al., 2017).

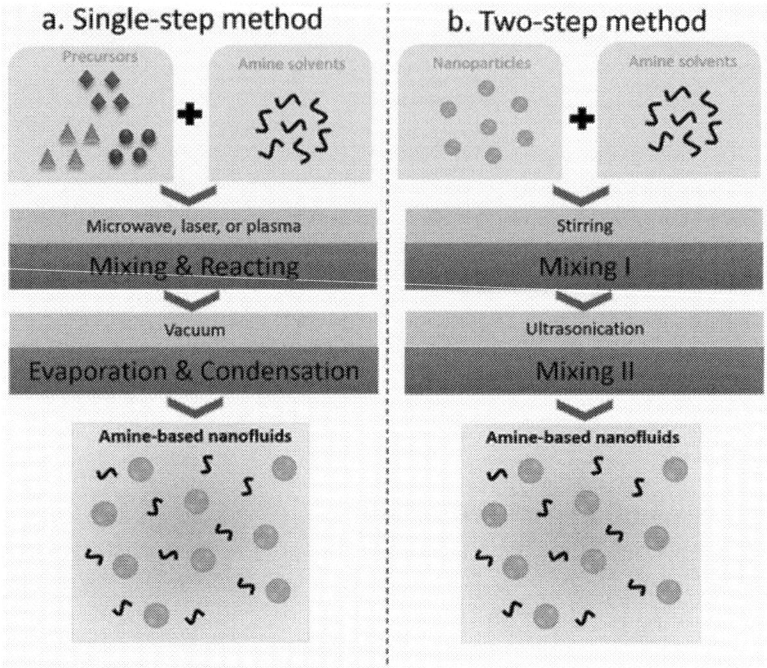

Figure 4.7 Single-step (A) and two-step (B) methods for the preparation of simple Nanofluids (Yu, Wang, Park, & Fang, 2019).

In general, nanofluids are produced through a single or two-step process (Fig. 4.7). Single step process is applied for small-scale production while the two-step process is used for large-scale production. The one-step process which is the simultaneous production of nanostructures and dispersion of them in the base fluid into a single step uses different methods. Direct evaporation is a one-step method that nanofluid is prepared using the solidification of the nanostructures, while they are initially in the gaseous phase inside the base fluid.

In the two-step process, the hybrid nanostructure is first prepared and then dispersed into the base fluid by means of a high shear mixing device. This process is very cheap for large-scale production, especially to produce nanofluids from available commercial nanostructures.

Stability is a key parameter which affects the performance of hybrid nanofluids. The lack of stability decreases nanofluid performance. Both of surface active and surface nonactive surfactants are used to reduce surface tension in fluid and facilitating the dispersibility of particles in the base fluid. It should be noted that the addition of excess quantity of surfactant into base fluid affects the viscosity, thermal conductivity, and stability of hybrid nanofluid.

Table 4.5 presents the preparation details of different nanofluids. However, it is not possible to compare the stability of nanofluids reported by different researchers because of the different conditions that affect the results. The stability is influenced by various parameters such as pH, nixing time, volume fraction, nanostructure size and type, base fluids, surfactants, and nanofluid synthesis procedure.

4.6 Capture of gases via sorption

The adsorbents are a prominent candidate for high-pressure compressed gas storage. Activated carbons and zeolites have used for the storage of natural gas at low pressures. Although the conventional adsorbents have not sufficient performance for gas storage, advanced materials such as modified activated carbons and metal-organic frameworks (MOFs) are known as alternative choices. The conventional adsorbents have small retention capabilities per adsorbent mass. Accordingly, there is widespread interest in the finding of novel adsorbent materials in order to intensification of the storage process.

Nanotechnology could help to introduce novel adsorbent or modifying existence adsorbents. The nanostructure adsorbents have a high surface to volume ratio, whereas they can be synthesized with specific physicochemical properties (Alonso et al., 2017). Table 4. 6 summarizes the most representative nanostructures for CO_2 and CH_4 sorption, while among the different adsorbents which are mentioned, Fe_3O_4-graphene and the MOF-117 based NPs have the greatest CO_2 sorption capacities. IRMOF-6, ordered mesoporous carbon, MOF-177 and MOF-5 have shown the highest adsorption capacities for CH_4 capture, compared to the graphene-based NPs, owing to their high thermal stability and the presence of high porosity in those structure.

4.7 Technoeconomic views on nanostructure as gas sorbents

The economic justification of gas capture performance using nanostructures is an undeniable factor in the large-scale application which should be evaluated. The cost of gas capture process includes cost of sorbents, cost of equipment installation and commissioning, operation and maintenance

Table 4.5 Summary of nanofluids preparation (Haddad, Abid, Oztop, & Mataoui, 2014).

Particle material	Base fluid	Parameters	Surfactant/pH control	Nanofluid stability
TiO_2 spherical	DI-water	$\Phi = 0.1-2$ vol.% $D = 15$ nm	Hexamethyldisilazane	Stable for several days without any visible sedimentation.
TiO_2 spherical	DI-water	$\Phi = 1.0$, 2.5 and 4.9 wt.% $D = 20$ nm	$pH = 11$	Stable for months
ZnO Approximately spherical	EG	$\Phi \leq 3$ vol.%	Ammonium citrate	Stable for several months
ZnO Nearly spherical	EG	$\Phi = 0.5-3.75$ vol.% $D < 50$ nm	–	Stable for 30 days without any trace of visible sedimentation
ZnO Nonspherical	Deionized water	$\Phi = 0.075$, 0.25 and 0.5 wt.% $D \sim 80$ nm	acac	Stable over 9 months to 1 year
SiO_2	TH66	$\Phi = 1.2-7$ vol.% $D = 15$ nm	BZC, CTAB, BAC	Nanofluids appeared stable without any Visual phase separation For at least a week
SiO_2	Deionized water	$\Phi = 0.2$, 0. 5 and 1 wt.% $D \sim 20$ nm	$pH = 4.5$	Nanofluids exhibited good stability over time
CuO	Oil	$\Phi = 0.2-2$ wt.% $D = 50$ nm	–	The complete sedimentation occurred After a week
CuO	DI-water	$\Phi = 0.1$ and 0.2 vol.% $D = 10-15$ nm	–	Very little settlement of nanoparticles even After 25 days

Al_2O_3	EG	$\Phi = 1$, 3 and 4 wt.% D = 20 nm	–	Dispersions were remained uniform for the duration of the Experiments
Al_2O_3	DI-water	$\Phi < 3$ vol.% D = 47 nm	SDBS	Nanofluid was observed to be stable For over a week
Fe_3O_4	Deionized water	$\Phi = 1-4$ vol.% D = 20–40 nm	PVP	Nanofluids were stable for about two weeks
Fe_3O_4	Kerosene	$\Phi = 0-1.0$ vol.% D = 15 nm	Oleic acid	Nanofluids were stable
CNTs	Double DI-water	$\Phi = 0.2$ wt.% D = 10–30 nm	Triton X-100	Nanofluid was stable up to 10 days
CNTs	EG	$\Phi = 0.2-1.0$ vol.%	NHS	The sedimentation of the suspensions was not visible
Cu	DI-water	$\Phi = 0.25$, 0.5 and 1.0 wt.% D = 10 nm	SDS	Nanoparticles were dispersed in water Evenly even after 10 h
Cu	R113	$\Phi = 0.25$, 0.5 and 1.0 wt.% D = 20 nm	SDS, CTAB and Span-80	Nanofluids were kept stable for 24 h

Table 4.6 Summary of characteristics of nanomaterial based sorbents for CO_2 and CH_4 (Alonso et al., 2017).

Sorbent	Maximum CO_2 capacity (mg CO_2/g sorbent)	Maximum CH_4 capacity (mg CH_4/g sorbent)	Experimental conditions (P,T)	Process scale	Key strengths	Main drawbacks
DD3R nanozeolite	123	26.6	3 MPa, 273–348K	Lab scale	• High selectivity for CO_2 • Used for biogas separation	• At pressures higher than 2000 kPa, high-pressure data points deviate from the MSL model-fit lines
T-type zeolite NPs	176.4	–	0.1 MPa, 288K	Lab scale	Potential application in the natural gas purification process • Recyclability • Applicability for postcombustion CO_2 separation	• N100 h to synthesize the T-type zeolite • Decrease of the adsorption capacities with increasing temperature from 288 to 313K
Zeolite NaA nanocrystals	215.6	–	0.1 MPa, 293K	Lab scale	• Use of a cost-effective method for the synthesis of highly active nanosized zeolites • High adsorption capacity for CO_2	• No expected dependence of the diffusion time constant on the square of the radius of the crystals for the CO_2 uptake

Ordered mesoporous carbon, SBA-15 Template	–	412	b7 MPa, 275K	Lab scale	Wet material has adsorption capacity 31% higher than AC	Long time to reach the equilibrium if the water content is higher than the critical value
Ni_2(dhtp) MOF	–	283.4	3.5 MPa, 298K	Lab scale	–	• Desorption approaches have not been adequately researched
Graphene	950.4	–	1.1 MPa, 298K	Lab scale	• High adsorption capacity compared to other carbon nanostructures • Possible use for industrial applications	• High pressures
Fe_3O_4-graphene	2640	–	1.1 MPa, 298K	Lab scale	• Easy synthesis • Can be reused	• High pressures
MW-CNTs@ JUC32	67.8	12	Patm, 273K	Lab scale	• 70% and 90% higher than nonmodified MW-CNTs • increase the CO_2 adsorption enthalpy	–

(Continued)

Table 4.6 Summary of characteristics of nanomaterial based sorbents for CO_2 and CH_4 (Alonso et al., 2017).—cont'd

Sorbent	Maximum CO_2 capacity (mg CO_2/g sorbent)	Maximum CH_4 capacity (mg CH_4/g sorbent)	Experimental conditions (P,T)	Process scale	Key strengths	Main drawbacks
IRMOF-6	–	409	3.500 MPa, 298K	Lab scale and simulation	–	• Chemical instability upon time. • High pressures
MOF-5	480	171.5	1.4 MPa for CO_2, 10 MPa for CH_4, 298K	Lab scale	• Easily tunable pore and chemical characteristics	• Lack of experimental data on performance after multiple cycles
MOF-177	1452	–	4.5 MPa, RT	Lab scale	• High thermal stability • Extra high porosity • High adsorption capacity at elevated pressures	• Negatively impacted by NO_x, SO_x, and H_2O • Low selectivity in CO_2/N_2 gas streams

costs, additional fuel resources, and gas compression, and the cost of transport and storage. A sustainable gas capture process should be operated at high capture capacity. Such a process should not only be able to use sorbent throughout multiple sorption/desorption cycles but also should have a low production cost. In this regard, to develop an economical process low-cost raw materials with simple synthesis production procedure should be used. The investigations for CO_2 capture show that the high price of the pertinent raw materials makes them unattractive compared to other metal salts. For example, the cost of lithium carbonate (Li_2CO_3) is $4.47 per kg, whereas the cost of $CaCO_3$ (raw material for CaO), at the same time, is only $0.005 per kg (Alonso et al., 2017). Economic evaluations for using the carbonaceous adsorbents indicate that CNTs with a specialized structure are only available at extremely high prices (i.e., $5/g) compared to granular activated carbon, which typically costs around $1/kg (Alonso et al., 2017).

Also, reusability of the captured CO_2 would also be an added advantage in order to make the CO_2 removal process economically feasible. Moreover, a CO_2 removal system that operates close to flue gas emission temperatures would be ideal as no extra energy would be needed. Based on the above, the challenges are to decrease the cost of capture and to scale up the technologies to the size required so that to minimize large-scale power generation emissions. For example, a large coal plant of 1000MW will typically generate about 8000 tons of CO_2 per day or about 3 million tons of CO_2 per year, provided that it runs at base load. This is more than one order of magnitude larger than the capacities of existing separation technologies. From the point of view of cost, the current technologies, if applied, would cost between €20 and €50 per ton of CO_2 separated, which would increase the cost of the electricity produced by 30% for a large coal plant. Fossil fuel based power generation with CO_2 capture and storage technologies would be competitive if these costs are properly addressed and reduced (Alonso et al., 2017).

4.8 Effective absorbance ratio

The enhancement percentage (Eq. 4.1) and effective absorption ratio (Eq. 4.2) are used to examine the absorption performance improvement by nanofluid compared to base fluid. Furthermore, these factors evaluate the effects of nanostructure on CO_2 absorption. The enhancement percentage is defined as the difference in the volume of CO_2 absorbed by the nanofluid and the base fluid divided by the volume absorbed by the base fluid, multiple by 100. The effective absorption ratio is defined as the volume of CO_2 absorbed by the nanofluid divided by the volume absorbed by the base fluid (Arshadi et al., 2019) (Table 4.7).

Table 4.7 Common use of nanofluids for CO_2 absorbance improvement (Arshadi et al., 2019).

E	Enhancement (%)	Base fluid	Nanoparticle	T (°C)	Solid loading	Contactor type
1.7	70.3	Water	$Fe_3O_4@SiO_2-SNH_2$	25	0.4W%	Bubble column
1.29	–	MDEA	TiO_2	35	0.06 kg/m^3	Bubble column
1.13	–	MDEA	Al_2O_3	35	0.08 kg/m^3	Bubble column
1.09	–	MEA	TiO_2	35	0.08 kg/m^3	Bubble column
1.04	–	MEA	Al_2O_3	35	0.06 kg/m^3	Bubble column
1.03	4.5	Methanol	Al_2O_3	20	0.01 vol%	Bubble column
1.025	–	Methanol	SiO_2	20	0.01 vol%	Bubble column
1.11	–	MDEA	TiO_2	20	0.8W%	Bubble column
–	24	Water	SiO_2	25	0.021W%	Bubble column
–	8.3	Methanol	Al_2O_3	10	0.01 vol%	Bubble column
–	33	DEA	Al_2O_3	25	0.05 vol%	Wetted wall column
–	40	DEA	SiO_2	25	0.05 vol%	Wetted wall column
–	9.4	Methanol	Al_2O_3	22	0.05 vol%	Tray column
–	9.7	Methanol	SiO_2	22	0.05 vol%	Tray column
–	21	Water	SiO_2	35	0.1W%	Batch vessel
–	18	Water	Al_2O_3	35	0.1W%	Batch vessel
–	25.9	Water	SiO_2	30	0.05W%	HFMC
–	3	Water	Al_2O_3	30	0.05W%	HFMC

$$\text{Enhancement}\ (\%) = \frac{V_{\text{Absorbed CO}_2\text{ by nanofluid}} - V_{\text{Absorbed CO}_2\text{ by base fluid}}}{V_{\text{Absorbed CO}_2\text{ by base fluid}}} \times 100 \tag{4.1}$$

$$\text{Effective absorbance ratio}(E) = \frac{V_{\text{Absorbed CO}_2\text{ by nanofluid}}}{V_{\text{Absorbed CO}_2\text{ by base fluid}}} \tag{4.2}$$

where V_{CO_2} is the total volume of CO_2 absorbed.

4.9 Mass transfer enhancement mechanisms

One of the main mechanisms that justify the effect of nanoparticle for enhancement of mass transfer, is the bubble breaking effect (Fig. 4.8). Based on this mechanism, during the bubble absorption process, the nanoparticles collide against each other and against the bubble. As the bubble rises up and the motion become more dynamic, the nanoparticles collide at the gas–liquid interface, and finally break the bubbles (Kim et al., 2014). Accordingly, the size of the bubbles become smaller leads to create a larger interfacial area and consequently causes intensification of mass transfer (Zhang et al., 2018).

Another involved mechanism is the grazing (or shuttle) effect which happens when the gas is transported by nanoparticles in the liquid (Fig. 4.9). Nanoparticles are rotated in the liquid while the gas is adsorbed by the

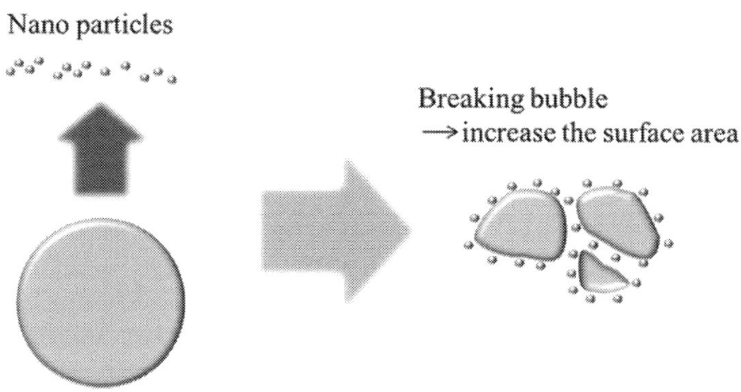

Figure 4.8 The schematic diagram of the bubble breaking effect (Kim, Jung, & Kang, 2014).

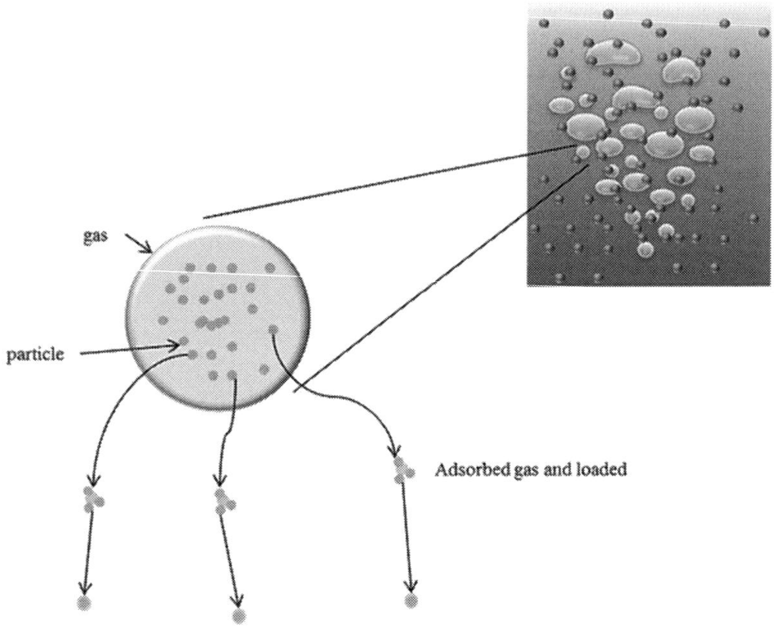

Figure 4.9 The schematic diagram of the shuttle effect (Kim, Jung, & Kang, 2014).

particles at the gas—liquid surface. The adsorbed gas is transferred from a high concentration area to a low concentration area due to the chemical potential gradient, while the nanoparticles move back to the primary place to adsorb the gas again. Under this condition, the surface and internal flow become more dynamic which cause an increase in the mass transfer rate (Kim, Jung, & Kang, 2014; Zhang et al., 2018). The addition of nanoparticles in a multiphase flow such as gas-liquid-solid system lead to promote the mass transfer performance due to strong adsorption of the diffusing gas-phase component in the dispersed phase particles. In fact, the concentration of the gas-phase reactant in the liquid phase near the interface will be decreased that improves the mass transfer rate. It is noteworthy that after spending a certain contact time, the particle will move back to the bulk where the gas-phase species is desorbed and the particles are regenerated. This phenomenon increases the mass transfer coefficient (Zhang et al., 2018).

The hydrodynamic effect is another mechanism involved in the adsorption using nanofluids (Fig. 4.10). Under the influence of this mechanism, the

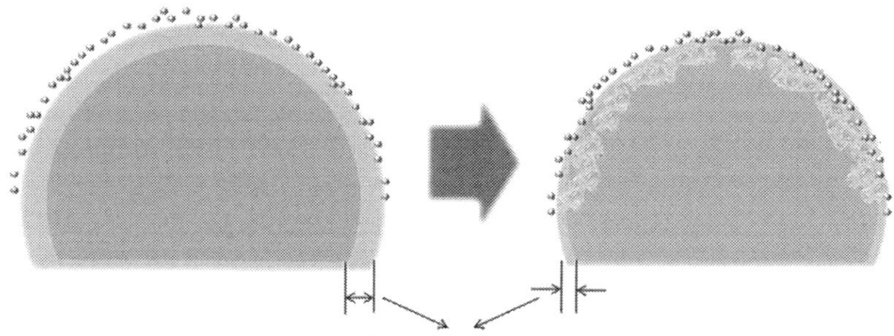

Figure 4.10 The schematic diagram of the hydrodynamic effect (Kim, Jung, & Kang, 2014).

diffusion boundary layer is broken by the nanoparticles surrounding the bubbles and convert to thin layers, whereas the diffusion rate into the liquid film enhances by the presence of the particles near the bubble-liquid interface (Kim, Jung, & Kang, 2014; Zhang et al., 2018). In fact, the hydrodynamic effect mechanism is the mixing of the gas—liquid boundary layer which causes a reduction in film layer thickness to produce thinner effective layers (Fig. 4.10). Gas diffusion into the liquid film increases by the presence of the particles near the bubble-liquid interface. Under this condition, the turbulence of the fluid flow increases which leads to enhances mass transfer coefficient. The reduction in the diffusion layer thickness owing to the collision of particles and interface improves the gas—liquid mass transfer rate, with the presence of inert and large particles (Zhang et al., 2018).

As two bubbles collide each other in a liquid medium, the film of liquid between the two gases drains over time to submicron thickness The balance between surface tension, hydrodynamic and surface forces affect the film thickness, while the very thin thickness will be ruptured which causes the bubble coalescence. The bubble coalescence in the gas—liquid dispersions will be inhibited with a mixed liquid phase. The small particles could alter the specific interfacial area and control the overall mass transfer coefficient. Generally, the gas absorption enhancement is influenced by more than one mechanism. It has been observed that in the gas absorption process by nanoemulsion absorbents, the absorption performance corresponds to the grazing effect and hydrodynamic effect (Zhang et al., 2018).

4.10 Economic evaluation/justification

Although the nanotechnology has great potential in drinking water treatment and pollution removal, for large-scale application the cost and environmental implications should be evaluated compared to the traditional processes. It seems that nanotechnology-based processes are generally are more expensive, but using available cheap raw materials they offer cheaper and more effective alternatives rather than conventional processes. The nanobased techniques could show undeniable performance in confronting with emerging pollutants, especially for the removal of low levels of contaminants to satisfy stringent water quality standards.

Table 4.8. indicates details of the removal of the contaminants using different processes based on specified time period, removal capacity, and cost. The cost of each process has estimated by cost calculation for treatment of volume of wastewater with a given technology relative to the typical influent concentration for given contaminants as the actual cost per mass of contaminants treated is scarcely measured. Findings indicate that for many contaminants, the nanobased treatment technologies not only are efficient but also create lower sludges compared to the traditional procedures. Furthermore, they are faster for operation in a wider range of aqueous chemistries. For instance, results for using halogenated and nonhalogenated organic reveal that both metal oxide nanoparticles and magnetic nanoparticles are comparable in effectiveness to the most effective traditional technologies such as carbon-based adsorption. The costs, however, are variable. In some cases, they are comparable in cost (e.g., nZVI used to treat cadmium, chromium, and zinc, and Mag-PCMA used to treat xylene and toluene are comparable in cost to current technologies) whereas in others the nanotechnology appears to be the less expensive option (nZVI used to treat TCE, lindane, chloramphenicol, lead, and arsenic and Mag-PCMA used to treat 2-chlorophenol, 589 naphthalene, methyl orange, atenolol, gemfibrozil, ethylbenzene, atrazine, and diuron; Adeleye et al., 2016). For emerging contaminants, more expensive traditional treatment process such as ozonation and reverse osmosis are often comparable in performance to the nanobased treatment processes.

Table 4.8 Data on the removal efficiency over a specified time period, removal capacity, and/or cost (Adeleye et al., 2016).

Treatment method and particle used	Pollutant	Removal efficiency (%/time)	Removal capacity (mg/g)	Removal cost ($/mg)
Activated carbon adsorption	TCE	83/2 wk	<12	0.28–1.46
Biodegradation + activated carbon	TCE	82.5/48 h	3.2	>0.006
Me/MeO (Fe/Ni coated membrane)	TCE	100/2 h	100	–
Activated carbon adsorption	Rhodamine B	93/5 h	6–16	–
Me/MeO (SiO_2/TiO_2 fibers)	Rhodamine B	100/2.7 h	300	
Carbon adsorption	Naphthalene	99/4 d	<450	0.21–19
Ion-exchange resin	Naphthalene	87/7 d	20	0.03–0.21
Magnetic (Mag-PCMAs)	Naphthalene	94/30 min	0.82	0.0048
Adsorption	Methylene blue	>99/2 h	7.5–714	<0.0033
Clay adsorption	Methylene blue	>99.9 h	300	0.066
Me/MeO (TiO_2 nanotubes)	Methylene blue	87/0.67 h	174	–
Ozone oxidation	Gemfibrozil	>76/4 min	–	0.01–60.00
Magnetic (Mag-PCMAs)	Gemfibrozil	57%	57	0.00004
Ion exchange	Copper	100	10–40	0.0001–0.004
Activated carbon adsorbents	Copper	–	83–150	0.0005–0.04
Carbon adsorption	Phosphate	>99	2.5–1150	0.00003–0.00004
Me/MeO (Mn/Al oxide)	Phosphate	100/5 h	60	–
Ion-exchange resin	Nitrate	100/12 h	14–82	0.0000001–0.001
Carbon adsorption	Nitrate	97/45 min	1–100	0.000001–0.01

4.11 Challenges and opportunities

Although nanostructures have many advantages and consist of relatively high capture capacities, reusability, and relatively low energy requirements, they suffer some drawbacks. Tuning and control of their size and dimensions during the synthesis process need costly materials, and in some cases, the synthesis process is complicated. In this regard, future research should focus on designing simple and low-cost production in large-scale. Furthermore, other technoeconomic considerations including storage capacity and the adsorption/desorption cycle durability should be improved to achieve a sustainable process.

For gas capture process using nanofluids, there are several serious challenges which should be overcome by rigorous research and design:

- The disagreement between most of the experimental findings.
- The lack of exact characterization of the nanofluids.
- The lack of complete understanding of the mechanism involved responsible for the behavior of nanostructures and nanofluids.

Moreover, the production of nanofluids faces some challenges in the fields of technical and financial. One of the main technical problems is the preparation of a homogeneous nanofluid, mostly because the particles tend to aggregate owing the strong van Der Waals interactions. The stability is a key factor for practical applications of nanofluids, while strongly influenced by the properties of the suspended particle and base fluid including the particle morphology, the chemical structure of the particles and base fluid. Accordingly, aggregation and sedimentation as two phenomena that affect the stability of nanofluids should be evaluated. The other existence issues of nanofluids are high production cost, anomalous high thermal conductivity, high viscosity, low specific heat and increased pressure drop as compared to the base fluid. Design and development of novel materials, equipment and processes can overcome remained challenges to achieve the commercialization of the nanofluids in many applications.

4.12 Conclusion and future perspectives

Although the nanostructures have many advantages for the removal of various pollutants, they still suffer from some drawbacks. The small size of nanostructures made it easy to aggregate which leads to blocking and fouling problems. Under this condition, the sorption capacity decreases, as well as secondary pollution will be produced. Some of the nanostructures have unstable surface properties and have great potential for oxidation. Another drawback of some nanostructures is the toxic effects related to the chemical synthesis methods. Furthermore, Large-scale application of nanostructures for wastewater treatment face a problem due to the biological incompatibility and high cost. In the case of metal ions, the low selectivity still remains as a serious issue. In this regards, the following solutions can be explored to overcome the limitations:

- Synthesis of nanostructures using green methods. One of the proposed source to produce nanostructures is the agricultural wastes, mineral slag, and clay materials. These sources are nontoxic and economical.
- Enhancement of the stability, selectivity, and sorption capacity by applying an efficient synthesis procedure, modification of surface properties, and subsequent geometric arrangement of nanostructures.
- Finding an efficient and eco-friendly process for regeneration and reusability of the adsorbents.

References

Adeleye, A. S., Conway, J. R., Garner, K., Huang, Y., Su, Y., & Keller, A. A. (2016). Engineered nanomaterials for water treatment and remediation: Costs, benefits, and applicability. *Chemical Engineering Journal, 286*, 640–662.

Alonso, A., Moral-Vico, J., Markeb, A. A., Busquets-Fité, M., Komilis, D., Puntes, V., et al. (2017). Critical review of existing nanomaterial adsorbents to capture carbon dioxide and methane. *The Science of the Total Environment, 595*, 51–62.

Araújo, C. S. T., Ione, L. S. A., Rezende, H. C., Marcionilio, S. M. L. O., Léon, J. J. L., & de Matos, T. N. (2018). Elucidation of mechanism involved in adsorption of Pb (II) onto lobeira fruit (Solanum lycocarpum) using Langmuir, Freundlich and Temkin isotherms. *Microchemical Journal, 137*, 348–354.

Arshadi, M., Taghvaei, H., Abdolmaleki, M. K., Lee, M., Eskandarloo, H., & Abbaspourrad, A. (2019). Carbon dioxide absorption in water/nanofluid by a symmetric amine-based nanodendritic adsorbent. *Applied Energy, 242*, 1562–1572.

Babu, J. A. R., Kumar, K. K., & Rao, S. S. (2017). State-of-art review on hybrid nanofluids. *Renewable and Sustainable Energy Reviews, 77*, 551–565.

Bharti, S., Thakur, L., Anand, S., & Dutta, V. (2020). Nano-based composites and their synthesis. In *Nanomaterials and environmental biotechnology*. Springer.

Bolisetty, S., Peydayesh, M., & Mezzenga, R. (2019). Sustainable technologies for water purification from heavy metals: Review and analysis. *Chemical Society Reviews, 48*, 463–487.

Burakov, A. E., Galunin, E. V., Irina, V. B., E Kucherova, A., Agarwal, S., Tkachev, A. G., et al. (2018). Adsorption of heavy metals on conventional and nanostructured materials for wastewater treatment purposes: A review. *Ecotoxicology and Environmental Safety, 148*, 702–712.

Chang, Y.-C., & Chen, D.-H. (2005). Adsorption kinetics and thermodynamics of acid dyes on a carboxymethylated chitosan-conjugated magnetic nano-adsorbent. *Macromolecular Bioscience, 5*, 254–261.

Chen, Y., Fan, Z., Zhang, Z., Niu, W., Li, C., Yang, N., et al. (2018b). Two-dimensional metal nanomaterials: Synthesis, properties, and applications. *Chemical Reviews, 118*, 6409–6455.

Chen, S., Li, R., Li, X., & Xie, J. (2018a). Electrospinning: An enabling nanotechnology platform for drug delivery and regenerative medicine. *Advanced Drug Delivery Reviews, 132*, 188–213.
Das, & Kumar, P. (2017). A review based on the effect and mechanism of thermal conductivity of normal nanofluids and hybrid nanofluids. *Journal of Molecular Liquids, 240*, 420–446.
Devendiran, D. K., & Amirtham, V. A. (2016). A review on preparation, characterization, properties and applications of nanofluids. *Renewable and Sustainable Energy Reviews, 60*, 21–40.
Duan, S., Gu, M., Du, X., & Xian, X. (2016). Adsorption equilibrium of CO_2 and CH_4 and their mixture on Sichuan Basin shale. *Energy and Fuels, 30*, 2248–2256.
Dubey, S., Banerjee, S., Nath Upadhyay, S., & Sharma, Y. C. (2017). Application of common nano-materials for removal of selected metallic species from water and wastewaters: A critical review. *Journal of Molecular Liquids, 240*, 656–677.
Fiyadh, S. S., AlSaadi, M. A., Jaafar, W. Z., Khalid Alomar, M., Fayaed, S. S., Mohd, N. S., et al. (2019). Review on heavy metal adsorption processes by carbon nanotubes. *Journal of Cleaner Production, 230*, 783–793.
Gamallo, M., Fernández, L., Feijoo, G., & Moreira, M. T. (2020). Nano-based technologies for environmental soil remediation. In *Nanomaterials for sustainable energy and environmental remediation*. Elsevier.
Ghasemi, M. H., Irani, V., & Ahmad, T. (2020). Amino functionalized ZIF-90@ GO/MDEA nanofluid: As a new class of multi-hybrid systems to enhance the performance of amine solutions in CO_2 absorption. *Journal of Natural Gas Science and Engineering, 74*, 103110.
Gómez-Pastora, J., Bringas, E., & Ortiz, I. (2014). Recent progress and future challenges on the use of high performance magnetic nano-adsorbents in environmental applications. *Chemical Engineering Journal, 256*, 187–204.
Gorji, T. B., & Ranjbar, A. A. (2017). A review on optical properties and application of nanofluids in direct absorption solar collectors (DASCs). *Renewable and Sustainable Energy Reviews, 72*, 10–32.
Haddad, Z., Abid, C., Oztop, H. F., & Mataoui, A. (2014). A review on how the researchers prepare their nanofluids. *International Journal of Thermal Sciences, 76*, 168–189.
Ha, S.-T., Su, R., Xing, J., Zhang, Q., & Xiong, Q. (2017). Metal halide perovskite nanomaterials: Synthesis and applications. *Chemical Science, 8*, 2522–2536.
Hayati, B., Maleki, A., Najafi, F., Gharibi, F., McKay, G., Gupta, V. K., et al. (2018). Heavy metal adsorption using PAMAM/CNT nanocomposite from aqueous solution in batch and continuous fixed bed systems. *Chemical Engineering Journal, 346*, 258–270.
Ibrahim, R. K., Hayyan, M., AlSaadi, M. A., Hayyan, A., & Ibrahim, S. (2016). Environmental application of nanotechnology: Air, soil, and water. *Environmental Science and Pollution Research, 23*, 13754–13788.
Istratie, R., Stoia, M., Păcurariu, C., & Locovei, C. (2019). Single and simultaneous adsorption of methyl orange and phenol onto magnetic iron oxide/carbon nanocomposites. *Arabian Journal of Chemistry, 12*, 3704–3722.
Jawed, A., Saxena, V., & Pandey, L. M. (2020). Engineered nanomaterials and their surface functionalization for the removal of heavy metals: A review. *Journal of Water Process Engineering, 33*, 101009.
Jin, X., Bobba, P., Reding, N., Song, Z., Thapa, P. S., Bala Subramaniam, G. P., et al. (2017). Kinetic modeling of carboxylation of propylene oxide to propylene carbonate using ion-exchange resin catalyst in a semi-batch slurry reactor. *Chemical Engineering Science, 168*, 189–203.
Khodadadi, H., Aghakhani, S., Majd, H., Kalbasi, R., Wongwises, S., & Afrand, M. (2018). A comprehensive review on rheological behavior of mono and hybrid nanofluids: Effective parameters and predictive correlations. *International Journal of Heat and Mass Transfer, 127*, 997–1012.

Kim, J. H., Jung, C. W., & Kang, Y. T. (2014). Mass transfer enhancement during CO2 absorption process in methanol/Al_2O_3 nanofluids. *International Journal of Heat and Mass Transfer, 76*, 484–491.

Koca, H. D., Doganay, S., Turgut, A., Tavman, I. H., Saidur, R., & Mahbubul, I. M. (2018). Effect of particle size on the viscosity of nanofluids: A review. *Renewable and Sustainable Energy Reviews, 82*, 1664–1674.

Lata, S., & Samadder, S. R. (2016). Removal of arsenic from water using nano adsorbents and challenges: A review. *Journal of Environmental Management, 166*, 387–406.

Lau, Y. J., Ahmed Khan, F. S., Mubarak, N. M., Lau, S. Y., Chua, H. B., Khalid, M., et al. (2019). Functionalized carbon nanomaterials for wastewater treatment. In *Industrial applications of nanomaterials*. Elsevier.

Liao, M.-H., Wu, K.-Y., & Chen, D.-H. (2005). Fast adsorption of crystal violet on polyacrylic acid-bound magnetic nanoparticles. *Separation Science and Technology, 39*, 1563–1575.

Lima, E. C. (2018). Removal of emerging contaminants from the environment by adsorption. *Ecotoxicology and Environmental Safety, 150*, 1–17.

Li, M., Sun, F., Shang, W., Zhang, X., Dong, W., Liu, T., et al. (2019). Theoretical studies of perfluorochemicals (PFCs) adsorption mechanism on the carbonaceous surface. *Chemosphere, 235*, 606–615.

Maity, J. P., Hsu, C.-M., Lin, T.-J., Lee, W.-C., Bhattacharya, P., Bundschuh, J., et al. (2018). Removal of fluoride from water through bacterial-surfactin mediated novel hydroxyapatite nanoparticle and its efficiency assessment: Adsorption isotherm, adsorption kinetic and adsorption thermodynamics. *Environmental Nanotechnology, Monitoring & Management, 9*, 18–28.

Mak, S.-Y., & Chen, D.-H. (2004). Fast adsorption of methylene blue on polyacrylic acid-bound iron oxide magnetic nanoparticles. *Dyes and Pigments, 61*, 93–98.

Mihoc, G., Ianoş, R., & Păcurariu, C. (2014). Adsorption of phenol and p-chlorophenol from aqueous solutions by magnetic nanopowder. *Water Science and Technology, 69*, 385–391.

Mondal, P., Anweshan, A., & Kumar Purkait, M. (2020). Green synthesis and environmental application of iron-based nanomaterials and nanocomposite: A review. *Chemosphere*, 127509.

Murgolo, S., Franz, S., Arab, H., Bestetti, M., Falletta, E., & Mascolo, G. (2019). Degradation of emerging organic pollutants in wastewater effluents by electrochemical photocatalysis on nanostructured TiO2 meshes. *Water Research, 164*, 114920.

Muthukumaran, C., Sivakumar, V. M., & Thirumarimurugan, M. (2016). Adsorption isotherms and kinetic studies of crystal violet dye removal from aqueous solution using surfactant modified magnetic nanoadsorbent. *Journal of the Taiwan Institute of Chemical Engineers, 63*, 354–362.

Navasingh, R. J. H., Kumar, R., Marimuthu, K., Planichamy, S., Khan, A., Asiri, A. M., et al. (2019). Graphene-based nano metal matrix composites: A review. In *Nanocarbon and its composites*. Elsevier.

Ni, B.-J., Huang, Q.-S., Wang, C., Ni, T.-Y., Sun, J., & Wei, W. (2019). Competitive adsorption of heavy metals in aqueous solution onto biochar derived from anaerobically digested sludge. *Chemosphere, 219*, 351–357.

Pashaei, H., & Ghaemi, A. (2020). CO2 absorption into aqueous diethanolamine solution with nano heavy metal oxide particles using stirrer bubble column: Hydrodynamics and mass transfer. *Journal of Environmental Chemical Engineering*, 104110.

Qadri, S., Ganoe, A., & Haik, Y. (2009). Removal and recovery of acridine orange from solutions by use of magnetic nanoparticles. *Journal of Hazardous Materials, 169*, 318–323.

Raghav, S., & Kumar, D. (2018). Adsorption equilibrium, kinetics, and thermodynamic studies of fluoride adsorbed by tetrametallic oxide adsorbent. *Journal of Chemical and Engineering Data, 63*, 1682–1697.

Rezakazemi, M., Darabi, M., Soroush, E., & Mesbah, M. (2019). CO_2 absorption enhancement by water-based nanofluids of CNT and SiO_2 using hollow-fiber membrane contactor. *Separation and Purification Technology, 210*, 920–926.

Saidi, M. (2020). CO_2 absorption intensification using novel DEAB amine-based nanofluids of CNT and SiO_2 in membrane contactor. *Chemical Engineering and Processing-Process Intensification, 149*, 107848.

Sajid, M. U., & Ali, H. M. (2019). Recent advances in application of nanofluids in heat transfer devices: A critical review. *Renewable and Sustainable Energy Reviews, 103*, 556–592.

Santhosh, C., Velmurugan, V., Jacob, G., Jeong, S. K., Grace, A. N., & Amit, B. (2016). Role of nanomaterials in water treatment applications: A review. *Chemical Engineering Journal, 306*, 1116–1137.

Shariati-Rad, M., Irandoust, M., Amri, S., Feyzi, M., & Ja'fari, F. (2014). Magnetic solid phase adsorption, preconcentration and determination of methyl orange in water samples using silica coated magnetic nanoparticles and central composite design. *International Nano Letters, 4*, 91–101.

Sheikh, M., Pazirofteh, M., Dehghani, M., Asghari, M., Rezakazemi, M., Valderrama, C., et al. (2019). Application of ZnO nanostructures in ceramic and polymeric membranes for water and wastewater technologies: A review. *Chemical Engineering Journal*, 123475.

Shen, C., Zhao, Y., Li, W., Yang, Y., Liu, R., & Morgen, D. (2019). Global profile of heavy metals and semimetals adsorption using drinking water treatment residual. *Chemical Engineering Journal, 372*, 1019–1027.

Sidik, N. A. C., Mahmud Jamil, M., Japar, W. M. A. A., & Adamu, I. M. (2017). A review on preparation methods, stability and applications of hybrid nanofluids. *Renewable and Sustainable Energy Reviews, 80*, 1112–1122.

Soares, S. F., Simões, T. R., António, M., Trindade, T., & Daniel-da-Silva, A. L. (2016). Hybrid nanoadsorbents for the magnetically assisted removal of metoprolol from water. *Chemical Engineering Journal, 302*, 560–569.

Sun, S., Zhang, X., Cui, J., Yang, Q., & Liang, S. (2019). High-index faceted metal oxide micro-/nanostructures: A review on their characterization, synthesis and applications. *Nanoscale, 11*, 15739–15762.

Taghipour, S., Mossa Hosseini, S., & Ataie-Ashtiani, B. (2019). Engineering nanomaterials for water and wastewater treatment: Review of classifications, properties and applications. *New Journal of Chemistry, 43*, 7902–7927.

Taheri, M., Mohebbi, A., Hashemipour, H., & Rashidi, A. M. (2016). Simultaneous absorption of carbon dioxide (CO_2) and hydrogen sulfide (H_2S) from $CO_2-H_2S-CH_4$ gas mixture using amine-based nanofluids in a wetted wall column. *Journal of Natural Gas Science and Engineering, 28*, 410–417.

Tan, X.-F., Liu, Y.-G., Gu, Y.-L., Xu, Y., Zeng, G.-M., Hu, X.-J., et al. (2016). Biochar-based nano-composites for the decontamination of wastewater: A review. *Bioresource Technology, 212*, 318–333.

Thangadurai, T. D., Manjubaashini, N., Thomas, S., & Maria, H. J. (2020). Nanostructured materials—design and approach. In *Nanostructured materials*. Springer.

Tran, H. N., Wang, Y.-F., You, S.-J., & Chao, H.-P. (2017). Insights into the mechanism of cationic dye adsorption on activated charcoal: The importance of $\pi-\pi$ interactions. *Process Safety and Environmental Protection, 107*, 168–180.

Wan, X., Zhang, X., Pan, W., Liu, B., Yu, L., Wang, H., et al. (2019). Ratiometric fluorescent quantification of the size-dependent cellular toxicity of silica nanoparticles. *Analytical Chemistry, 91*, 6088–6096.

Xu, J., Cao, Z., Zhang, Y., Yuan, Z., Lou, Z., Xu, X., et al. (2018). A review of functionalized carbon nanotubes and graphene for heavy metal adsorption from water: Preparation, application, and mechanism. *Chemosphere, 195*, 351–364.

Yang, X., Wan, Y., Zheng, Y., He, F., Yu, Z., Huang, J., et al. (2019). Surface functional groups of carbon-based adsorbents and their roles in the removal of heavy metals from aqueous solutions: A critical review. *Chemical Engineering Journal, 366*, 608–621.

Yu, W., Wang, T., Park, A.-H. A., & Fang, M. (2019). Review of liquid nano-absorbents for enhanced CO_2 capture. *Nanoscale, 11*, 17137–17156.

Zargar, B., H Parham, & Hatamie, A. (2009). Fast removal and recovery of amaranth by modified iron oxide magnetic nanoparticles. *Chemosphere, 76*, 554–557.

Zhang, Z., Cai, J., Chen, F., Li, H., Zhang, W., & Qi, W. (2018). Progress in enhancement of CO2 absorption by nanofluids: A mini review of mechanisms and current status. *Renewable Energy, 118*, 527–535.

Zhang, Y., Wu, B., Xu, H., Liu, H., Wang, M., He, Y., et al. (2016). Nanomaterials-enabled water and wastewater treatment. *NanoImpact, 3*, 22–39.

Zhua, H.-Y., Ru, J., Xiaob, L., & Lib, W. (2010). A novel magnetically separable-Fe_2O_3/crosslinked chitosan adsorbent: Preparation, characterization and adsorption application for removal of hazardous azo dye. *Journal of Hazardous Materials, 179*, 251–257.

Zhu, Y., Fan, W., Zhou, T., & Li, X. (2019). Removal of chelated heavy metals from aqueous solution: A review of current methods and mechanisms. *The Science of the Total Environment, 678*, 253–266.

Zhu, H. Y., Jiang, R., Fu, Y.-Q., Jiang, J.-H., Xiao, L., & M Zeng, G.- (2011). Preparation, characterization and dye adsorption properties of γ-Fe2O3/SiO2/chitosan composite. *Applied Surface Science, 258*, 1337–1344.

Zhu, K., Ju, Y., Xu, J., Yang, Z., Gao, S., & Hou, Y. (2018). Magnetic nanomaterials: Chemical design, synthesis, and potential applications. *Accounts of Chemical Research, 51*, 404–413.

CHAPTER FIVE

Membrane-based sorption processes

5.1 Introduction

Membranes have very important and various applications in chemical technology. Membranes are used to control the rate of permeation of a certain chemical species between two separated phases, e.g., the purpose of drug delivery systems is the control of realizing the drug from its source to the human's body, or in the separation systems, only one of the components is allowed to pass through membrane and the others are not allowed. The membranes differ in terms of chemical and physical composition and how they operate. Membranes adsorption do the doubled action of filtration and adsorption to remove small amounts of contaminants such as cationic heavy metals, anionic phosphates, and nitrates (Khulbe & Matsuura, 2018). Adsorption is one of the most effective mechanisms for removing compounds by membrane filtration in addition to size removal and charge desorption (Comerton, Andrews, Bagley, & Yang, 2007). Adsorption is a mass transfer process whereby a material is transferred from the liquid phase to the surface of a solid and bonded to physical and/or chemical interactions (Khulbe & Matsuura, 2018). Adsorption is awaited to be moved by hydrophobic and hydrogen bonding (Comerton et al., 2007). Membranes for industrial separation can be broadly classified into the following four groups according to the driving force (Nath, 2017):

- Pressure driven membrane process: reverse osmosis (RO), nanofiltration (NF), ultrafiltration (UF), microfiltration (MF), pervaporation (PV), and membrane gas separation.
- Concentration gradient drove membrane process: dialysis, membrane extraction.
- Electrical potential drove membrane process: electrodialysis (ED).
- Other membrane processes: facilitated or carrier mediated membrane transport, liquid membrane separation, membrane reactors, membrane distillation.

Intensification of Sorption Processes
ISBN: 978-0-12-821411-4
https://doi.org/10.1016/B978-0-12-821411-4.00006-7

The materials used to make the membrane should have the following main properties in order to be able to perform their separation function well (Nath, 2017):

- Chemical resistance to feed or materials used to clean membranes.
- Mechanical stability.
- Thermal stability.
- Permeability and high selectivity

In general, materials used to make membranes can be divided into three categories (Porter, 1989):

- Synthetic polymers such as polyamides, polysulfones, silicon rubber, and perfluoro polymers, which are one of the most important synthesized polymer materials for making membranes.
- Synthetic natural substances, such as cellulose-based materials.
- Other materials such as minerals, ceramics, and metals.

With regard to the type of operating process and the cost of investment, the suitable material for the membrane can be selected by the process.

Table 5.1 shows a list of common polymers used to make membranes from them.

Membranes can be structurally homogeneous or heterogeneous. The first type, the composition, and structure are completely uniform. However, the second type of membranes has a layered or porous structure. Membranes

Table 5.1 Membranes and their applications.

Membrane type	Applications
Polyethylene (PE)	Desalination, reverse osmosis
Poly(dimethylsiloxane) (PDMS)	Adsorption, gas separation
Polysulfone (PSU)	Forward osmosis, microfiltration
Polyethersulfone (PES)	Wastewater treatment, ultrafiltration
Polyimide (PI)	Gas separation, reverse osmosis
Polycarbonate	Gas separation, microfiltration
Polyethers	Reverse osmosis, gas separation
Polypyrrolones	Gas separation
Polysulfones	Wastewater treatment, membrane distillation
Polyethersulfones	Gas separation, ultrafiltration

From: Nath, K. 2017. Membrane separation processes (PHI Learning Pvt. Ltd.).

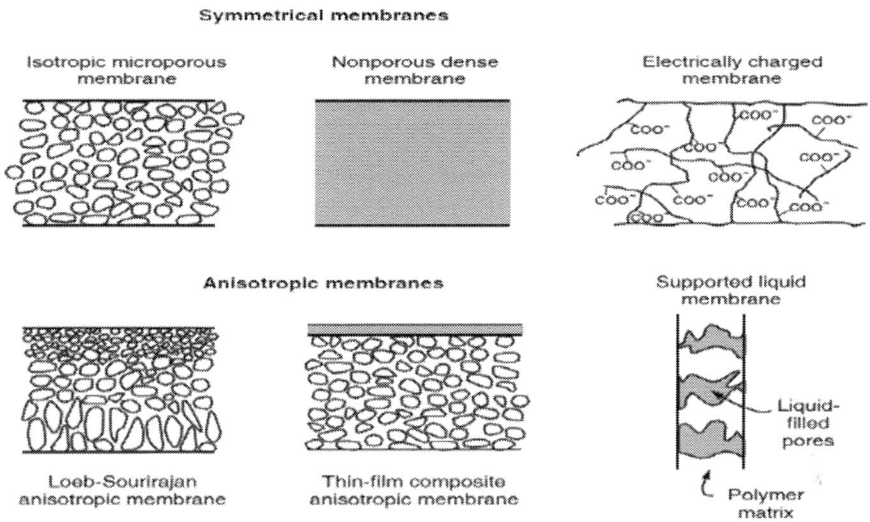

Figure 5.1 Schematic diagram of different types of membranes (Baker, 2004).

have several main types, such as isotropic, nonisotropic, ceramic, metal, and liquid membranes. Fig. 5.1 shows the principal types of membrane schematically (Khulbe & Matsuura, 2018).

For practical applications, membranes are installed in a suitable device, which is referred to as a membrane module. Different types of membrane modules are available (Fig. 5.2). There are basically four different designs of membrane modules which are most commonly used (Nath, 2017). These are:

(a) Plate and frame
(b) Tubular
(c) Spiral wound
(d) Hollow fiber

The membrane adsorbent is manufactured by attaching the functional groups to the surface and the pore wall of the polymer membranes. Pollutants are selectively absorbed into the functional group. The membrane absorber merges membrane filtration performance effectively. As the membrane passes through the membrane, the functional active sites are combined with the pollutants in question, absorbing them to remove contaminants from the drinking water at high absorption rate and capacity (see Fig. 5.3) (Khulbe & Matsuura, 2018).

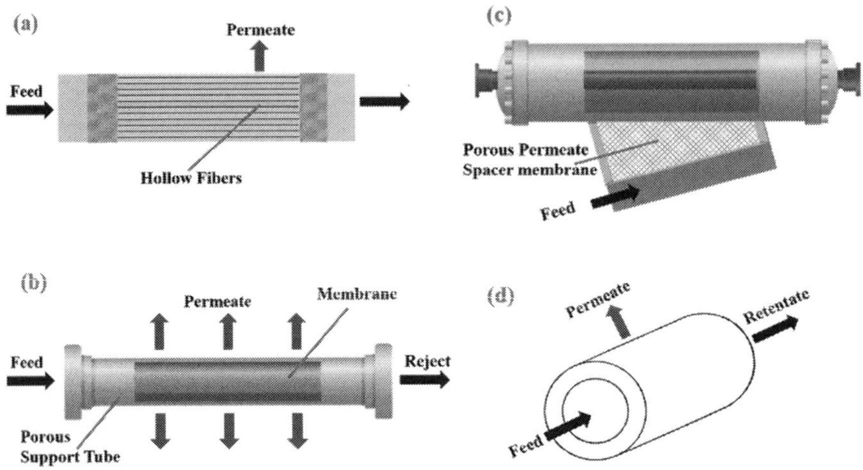

Figure 5.2 Schematic representation of various membrane modules: (A) plate and frame; (B) tubular; (C) spiral wound; (D) and hollow fiber. *From: Nath, K. 2017. Membrane separation processes (PHI Learning Pvt. Ltd.).*

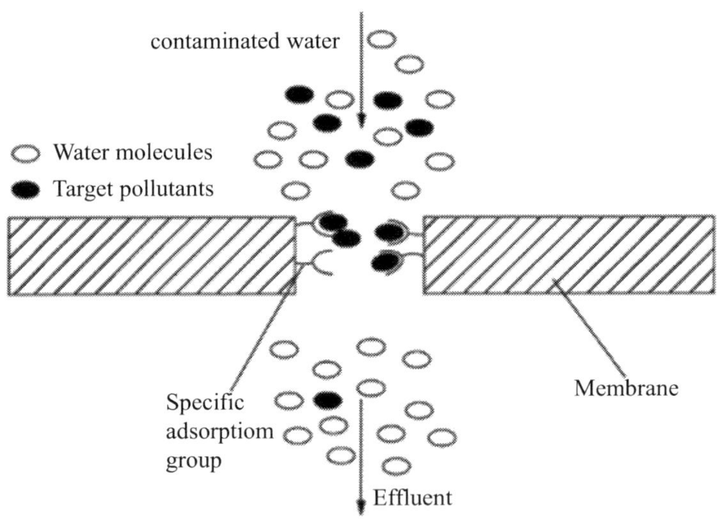

Figure 5.3 The principle of membrane adsorbent (Khulbe & Matsuura, 2018).

5.2 Membrane technologies

Some of the membrane technologies have been well studied and established. This group of technologies is widely used in various industries.

The processes in this category include microfiltration, ultrafiltration, reverse osmosis, and electrodialysis. Table 5.2 provides a summary of the industrial applications of membrane processes (Bai et al., 2015).

Membrane processes can be classified according to the driving force used in the process. A summary of the main characteristics of membrane processes is presented in Table 5.3 (Bai, Zan, Zhang, & Sun, 2015).

The specific aspect for the membrane processes that make them absorbing for industrial usage is their compactness, facility of making, function, and modular plan. Further benefits of membrane process cleanliness with low energy consumption, ease of use that can be easily used elsewhere in conventional processes can be recovering valuable products, and that

Table 5.2 Some industrial applications of membrane technology.

Membrane-based technology	Industrial application
Microfiltration, ultrafiltration, nanofiltration, reverse osmosis	Food and beverage
Membrane distillation	Petroleum refining
Nanofiltration	Paint, adhesive, and solvent recovery
Ultrafiltration	Plating processes and waste reduction
Ultrafiltration, nanofiltration, reverse osmosis	Water purification
Microfiltration, ultrafiltration, nanofiltration, reverse osmosis	Wastewater treatment

From: Nath, K. 2017. Membrane separation processes (PHI Learning Pvt. Ltd.).

Table 5.3 Characteristics of the membrane process.

Driving force	Process
Chemical potential gradient ($\Delta\mu$)	Vapor permeation
	Pervaporation
	Dialysis
	Gas separation
Pressure gradient (ΔP)	Reverse osmosis (7−75 bar)
	Nanofiltration (2−40 bar)
	Ultrafiltration (0.5−10 bar)
	Microfiltration (0.1−3 bar)
Electrical potential gradient (ΔE)	Electrodialysis
	Membrane electrolysis
Temperature gradient (ΔT)	Membrane distillation

From: Nath, K. 2017. Membrane separation processes (PHI Learning Pvt. Ltd.).

membrane filtration can contribute to the removal of the desired contamination by adsorption and increase the processing speed (Bai et al., 2015). The flux through the membrane can be generally expressed as:

$$Flux = \frac{membrane\ permeability}{(membrane\ thickness)\ (Driving\ force)} \tag{5.1}$$

This driving force may be a difference in pressure, concentration, temperature, chemical potential, and so on (Nath, 2017).

5.2.1 Ultrafiltration process

The Ultrafiltration process is used to separate water and microsolutes from macromolecules and colloids by a porous medium as a membrane. Solvents and salts of low molecular weight will pass through the membranes while larger molecules are retained by adsorbing with the surface and wall of the membrane and they are sieved with it (Fig. 5.4) (Scott & Hughes, 2012). The average pore diameter is between 10 and 1000 Å. Anisotropic structure of the ultrafiltration membrane is usually formed by a process so-called Loeb and Sourirajan process. Ultrafiltration membranes are made of a porous surface layer that is placed on microporous support. The porous surface layer is responsible for the separation, and the microporous bed plays a mechanical preservative role.

In 1861, Schmidt successfully completed the ultrafiltration experiment with a natural bovine pericardium for the first time in the laboratory (Li, Jiang, & Li, 2018). The first ultrafiltration membranes were made using nitrocellulose (Bechhold, 1907). Such membranes are made up to 1920, have laboratory use only. This continued until 1960 when in 1963 an asymmetric ultrafiltration membrane with acetate cellulose was constructed (Baker, 2004). The main objective goal was to produce high-flux reverse osmosis membranes, but other researchers realized the general applicability of the technique (Baker, 2004). Michaels (1971) produced ultrafiltration membranes from cellulose acetate and many other polymers including

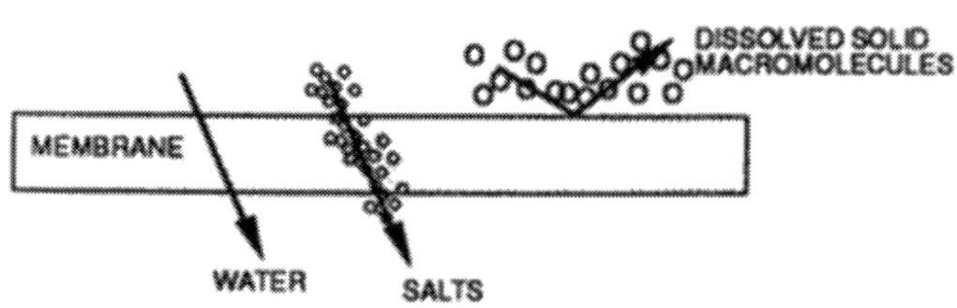

Figure 5.4 Separation by ultrafiltration (Scott & Hughes, 2012).

polyacrylonitrile copolymers, aromatic polyamides, polysulfone and poly (vinylidene fluoride). According to the reports (Li et al., 2018), plenty of water has been generated with integrated water treatment for the Yangtze River water treatment, with a flow rate of 16.83–17.67 L per minute, an outlet flow of 15.00–16.40 L per minute. Equivalent to 89%–93% penetration of water flow, water quality is good and the total area of 1.28 m^2 covers, which saves considerable space to the traditional process. A case study (Li et al., 2018) shows that UF's domestic water production cost is 0.23 Yuan/m^3, while the electricity consumption is 0.18 kW/m^3. The research of the Shenzhen water group shows that adding ozone and activated carbon advanced treatment technology on the basis of the conventional process will increase the water production cost by 0.3 Yuan/m^3. These studies show that the water cost of the ultrafiltration process is lower than that of the conventional process. A summary of the ultrafiltration process is given in Table 5.4.

There are several factors that can affect the performance of an ultrafiltration system, these factors are transmembrane flow, operating pressure, operating temperature, salt concentration (Nath, 2017).

Transmembrane flow: The rate of permeation increases with the increases in the flow velocity of the liquid across the membrane surface. Operating pressure: Permeation rate is directly proportional to the applied pressure across the membrane surface, due to increased deposition due to contaminant adsorption and density, the operating pressures rarely exceed

Table 5.4 Characteristics of the ultrafiltration process.

Characteristic	Range/description
Membrane type	Hollow fiber Polyacrylonitrile PVDF electrocoat Polyethersulfone Composite single layer Tubular
Driving force	Pressure gradient
Transport mechanism	Pore flow
Separation principle	Screening
Allowing retention	Compounds with a molecular weight of 300,000–500,000 Da
Pore size	1–100 nm

From: Nath, K. 2017. Membrane separation processes (PHI Learning Pvt. Ltd.).

100 PSIG and are generally around 50 psi. Operating temperature: Rate of permeation increases with increasing temperature. Salt concentration: The UF rate decreases with increasing molar concentration. The main problem of ultrafiltration is the flux decrease caused by concentration polarization and fouling. Concentration polarization causes a rapid drop in flux while fouling causes a gradual, long-term decay. The two flux-decline phenomena are illustrated in Fig. 5.5 (Jönsson & Trägårdh, 1990).

The "initial" solution flux is typically 5%—95% of the pure water flux (Jönsson & Trägårdh, 1990). The unrivaled ultrafiltration capability is to separate, purify, and concentrate aim macromolecules (Nath, 2017). Some ultrafiltration applications are presented below: Water treatment: Ultrafiltration membranes are effective at removing colloids, proteins and bacteria and other coarse organic molecules by removing larger molecules than the size of the cavities and trapping some of them by adsorption onto the surface. Therefore, ultrafiltration can be used in the pretreatment of the reverse osmosis process or the final process of deionized water (Nath, 2017). Clarification of juice: The ultrafiltration process eliminates opaque solids by adsorbing and separating the size while allowing water, sugar and the taste to pass through the membrane (Nath, 2017). Other applications are processing of the whey, concentration of latex emulsion, pulp and paper industry, etc. (Nath, 2017). The following is another application of ultrafiltration membranes: Because conventional UF membranes are highly permeable but cannot remove nitrate and phosphate from water, Qiang Gao and

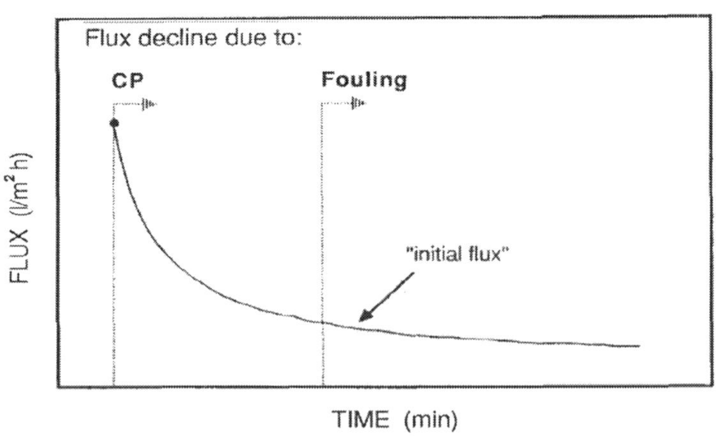

Figure 5.5 Flux decline is caused by concentration polarization (CP) and fouling. The pure water flux is indicated by (·) (Jönsson & Trägårdh, 1990).

colleagues first developed a UF membrane microreactor (MMR) to simultaneously remove nitrate and phosphate from the water with normal UF membrane permeability. UF MMR is a combination of zirconium hydroxide and quaternary ammonium powder with polyvinylidene fluoride (PVDF) and has the typical pore structure and size of UF membranes. In this process, the dead-end filtration method, a membrane with a diameter of 63.5 mm, and an effective area of 28.7 cm^2 have been used. The process is performed at a membrane pressure of 0.1 MPa supplied by nitrogen and the agitator speed is 250 rpm. The schematic of this process is shown in Fig. 5.6 (Gao et al., 2019).

As shown in Fig. 5.7, by ion-exchange, adsorption, and replacement of surface hydroxyl introduced into the UF membrane by zirconium hydroxide and SDMOAC, the generated microreactor successfully removed nitrate and phosphate from the water during the UF process (Gao et al., 2019).

Due to the favorable adsorption of nitrate and phosphate, which leads to the removal with high permeability, the described UF MMR is a promising prospect for advanced water and wastewater treatment (Gao et al., 2019).

5.2.2 Microfiltration

Microfiltration is a process in which porous membranes are used. This process is used to remove suspended particles from 0.1 to 10 μm in diameter. Therefore, the process of microfiltration membranes is a process between ultrafiltration membranes and conventional filters. Ultrafiltration and microfiltration are processes that are very similar and the difference between them

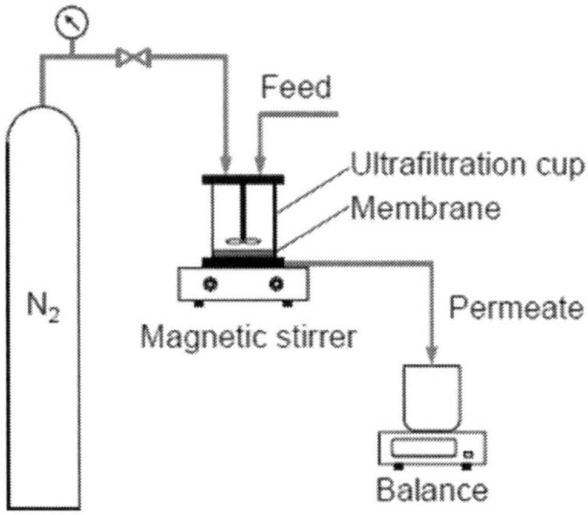

Figure 5.6 Schematic diagram of the filtration device (Gao et al., 2019).

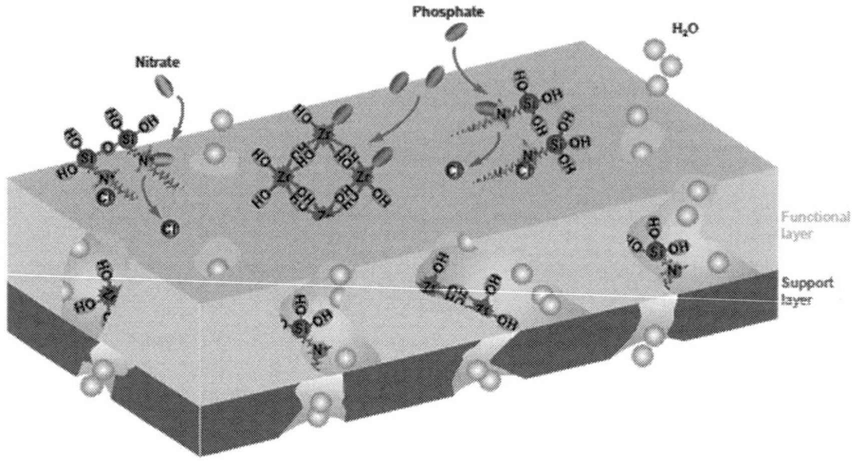

Figure 5.7 Schematic representation of the mechanism of nitrate and phosphate removal by UF MMR (Gao et al., 2019).

is the size of the membrane pores. Like ultrafiltration, microfiltration has its modern origins in the development of collodion (nitrocellulose) membranes in the 1920s and 1930s. In 1926 Membrane filter GmbH was founded and began to produce collodion microfiltration membranes commercially. The market was very small, but by the 1940s other companies, including Sartorius and Schleicher and Schuell, were producing similar membrane filters (Baker, 2004). The first application of microfiltration membranes has been the cultivation of microorganisms in drinking water, which has been used during the Second World War as a rapid method for the detection and monitoring of water source pollution (Nath, 2017). Until the mid-1960s, the use of microfiltration membranes was confined to laboratory or to very small-scale industrial applications. The introduction of pleated membrane cartridges by Gelman in the 1970s was an important step forward and made possible the use of microfiltration membranes in large-scale industrial applications. In the 1960s and 1970s, microfiltration became important in biological and pharmaceutical manufacturing, as did microfiltration of air and water in the production of microelectronics in the 1980s (Baker, 2004).

There are two main schemes in the microfiltration process (Fig. 5.8):

- Dead-end or in-line filtration.
- Cross-Flow Microfiltration.

As you can see in Fig. 5.8, in the cross-flow microfiltration (CFM), the feed is pumped tangentially onto the membrane and in the dead-end

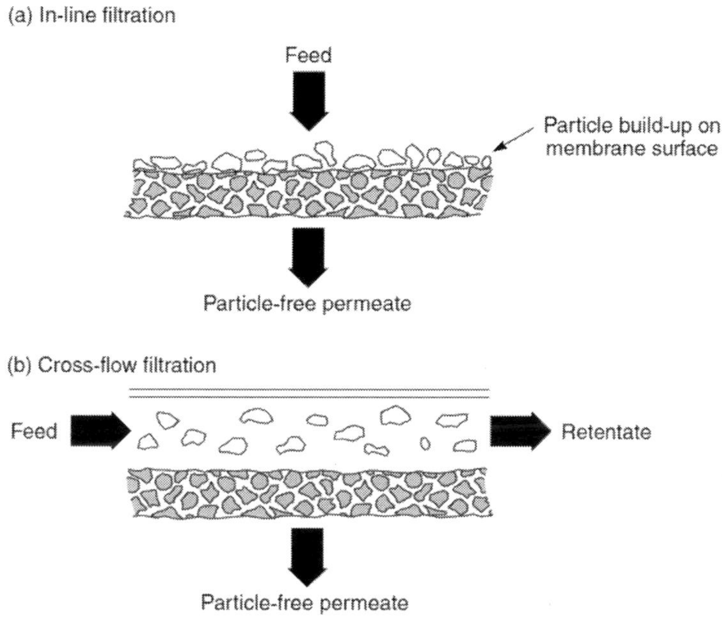

Figure 5.8 Schematic representation of (A) in-line and (B) cross-flow filtration with microfiltration membranes (Baker, 2004).

or in-line filtration arrangement, the feed flow is normal to the membrane surface (Nath, 2017). A summary of the microfiltration process is given in Table 5.5.

The advantages and disadvantages of in-line microfiltration and cross-flow filtration is compared in Table 5.6.

The microfiltration membrane can be divided into two categories based on their pore structure. These are depth membranes and screen membranes (Nath, 2017). Depth membranes: This is the most common type of

Table 5.5 Characteristics of the microfiltration.

Characteristic	**Range/description**
Driving force	Pressure gradient
Membrane structure	Porous, symmetric, asymmetric
Pore size	0.1−10 μm
Operation modes	Dead-end (in line) filtration Cross-flow (tangential flow) filtration
Separation mechanism	Sieving (molecular size)

From: Nath, K. 2017. Membrane separation processes (PHI Learning Pvt. Ltd.).

Table 5.6 Comparison of advantages and disadvantages of in-line and cross-flow microfiltration (Baker 2004).

Dead-end or in-line microfiltration	Cross-flow microfiltration
Low capital cost	High capital cost
High operating costs—membrane must be replaced after each use and disposal can be a problem	Operating costs modest—membranes have extended lifetimes if regularly cleaned
Simple operation—Parts do not move	Complex operation—Filters need regular cleaning
Best suited to dilute (low solid content) solutions. The cost of membrane replacement increases with the concentration of particles in the feed solution	Best suited to high solid content solutions. The cost is relatively independent of the concentration of soluble feed particles
Representative applications: Sterile filtration Clarification/sterilization of beer and wine	Representative applications: Continuous culture/cell recycle Filtration of oilfield produced water

microfiltration membrane. They are membranes with tortuous type pores. Depth membranes have relatively large cavities on their upper surface so that the particles pass through the upper surface of the membrane and finally stop in the membrane's internal structure or by absorbing on the inner walls of the membrane (Nath, 2017). Screen membranes: In screen membranes, the pores are not tortuous but straight through cylindrical pores. Screen membranes have small holes at their upper surface, particles with a diameter larger than these cavities accumulate and sometimes adsorbed on the membrane surface (Nath, 2017). In Fig. 5.9, the two main types of microfiltration membranes are shown.

Fouling is one of the problems that may occur in the microfiltration processes. Fouling in microfiltration membranes can be categorized into two main types, external fouling and internal fouling (Nath, 2017). External fouling or concentration polarization is a phenomenon in which a higher concentration of particle or molecules is retained on a membrane surface compared to that of the bulk of the feed stream. Internal fouling or pore blocking is the next step after concentration polarization. Due to concentration polarization, the interaction between retained components, and membrane surface occurs and components adsorbing on the surface, which can lead to irreversible fouling of the membrane (Fig. 5.10).

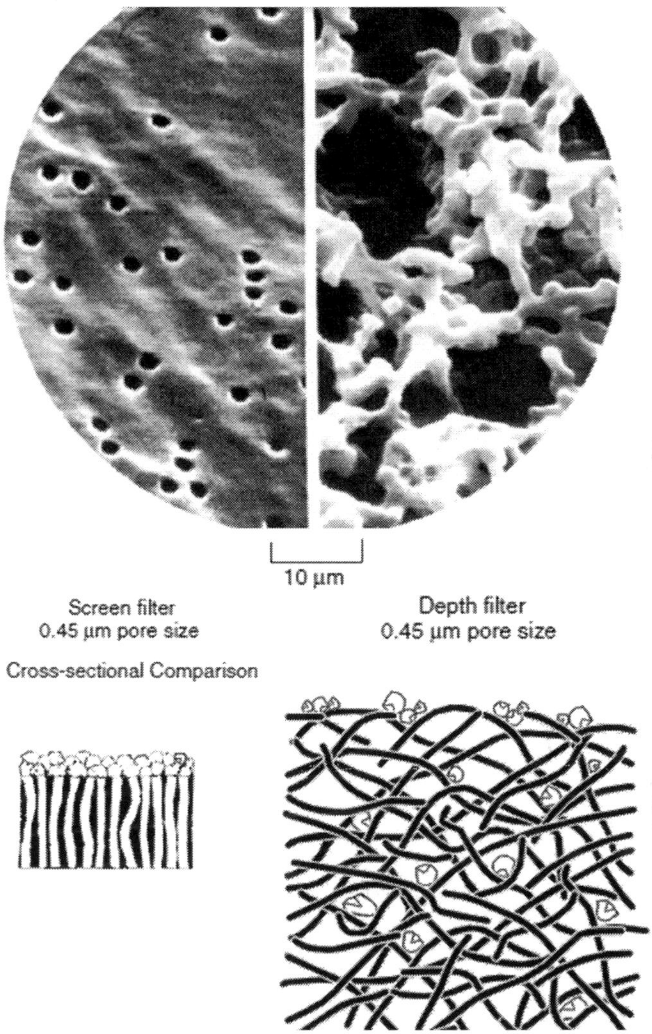

Figure 5.9 Surface scanning electron micrograph and schematic comparison of nominal 0.45-μm screen and depth filters. The screen filter holes are equal and tiny and take the maintained particles on the membrane surface (Baker, 2004).

Screen filters are quickly blocked by the accumulation and adsorbing of residual particles on the membrane surface, while depth filters, due to the fact that they have a larger area for particle accumulation and sorption, become stuck in a longer period of time. Some microfiltration applications are presented below: Recovery of cheese brine: cheese brine is produced during the production of cheese with various components such as cheese

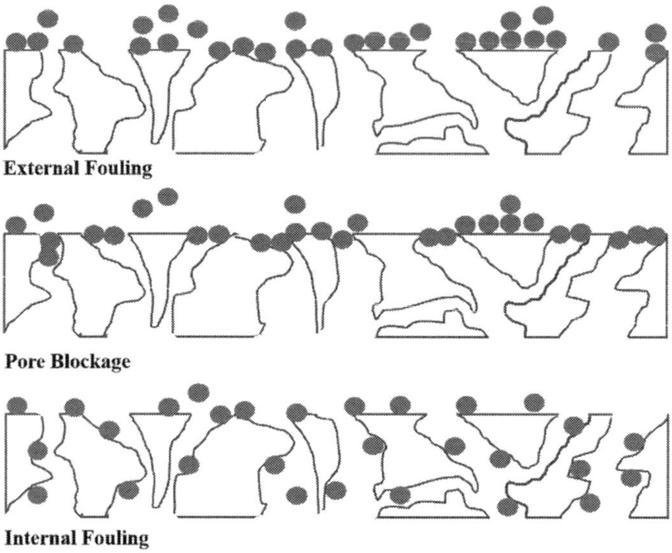

Figure 5.10 Schematic diagram of internal and external fouling. *From: Nath, K. 2017. Membrane separation processes (PHI Learning Pvt. Ltd.).*

particles, protein, fat, salt, lactose, and bacteria. Microfiltration reduces these particles and pollutants by adsorbing onto the surface and walls and excreting them. It also eliminates other pollutants such as foam, scams, etc. (Nath, 2017). Clarification of fermentation broth: In fermented broth, there is usually cell mass, suspended particulate matter and some protein. Microfiltration by sieving and adsorption can remove suspended solids from the fermentation fluid and make the fermentation liquid clear (Nath, 2017).

Other applications are Processing of ultrapure water, Pretreatment of whey, milk microfiltration, wine filtration, and industrial wastewater treatment. Another example of microfiltration (MF) membrane application is the rapid removal of anionic dyes from water by diffusion adsorption, which is briefly described below. For this process, a new porous composite membrane produced by placing a polyethyleneimine (PEI) polymer on an MF membrane has been used to remove anionic dyes from water by rapid absorption. Experiments with different dye solutions have shown that the anion dye adsorption capacity of this composite membrane in an acidic solution is high during rapid penetration. Fig. 5.11 shows the adsorption of dye on the membrane surface (Liu, Cheng, Zhao, & Zhu, 2017).

After decolorization, this membrane can be regenerated with a simple wash (Liu et al., 2017).

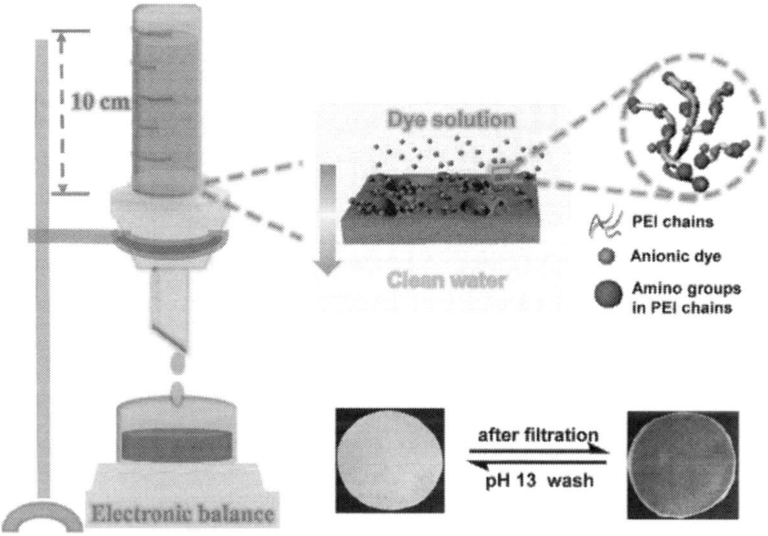

Figure 5.11 Schematic of the decolorization process (Liu et al., 2017).

5.2.2.1 Reverse osmosis

As the name suggests, reverse osmosis is osmosis in reverse. Therefore, we should understand the phenomenon of osmosis at first. Osmosis may be defined as the physical movement of a solvent through a semipermeable membrane based on a difference in chemical potential between two solutions separated by that semipermeable membrane (Nath, 2017). In conventional osmosis, the solvent is naturally transmitted through the membrane from the part where the water potential is high (High salt concentration) to the part where the water potential is low (low salt concentration). Water diffusion will continue until the solutions are substantially balanced or the rising brine column causes sufficient hydrostatic pressure to limit further diffusion. Using the external pressure to reverse the normal flow of pure solvent is called reverse osmosis. Show the osmosis process and reverse osmosis in Figs. 5.12 and 5.13, respectively.

Reverse osmosis is a process used to water desalting using membranes. In this process, water penetrates, but the salt is not able to penetrate into the membrane. On the side of the membrane where the feed is introduced, there is pressurized water with dissolved salt. By applying a higher hydraulic pressure than the sum of the osmotic pressure difference and the pressure loss of diffusion through the membrane, we can transfer the water in the opposite direction (Nath, 2017) and the water is separated from the salt and

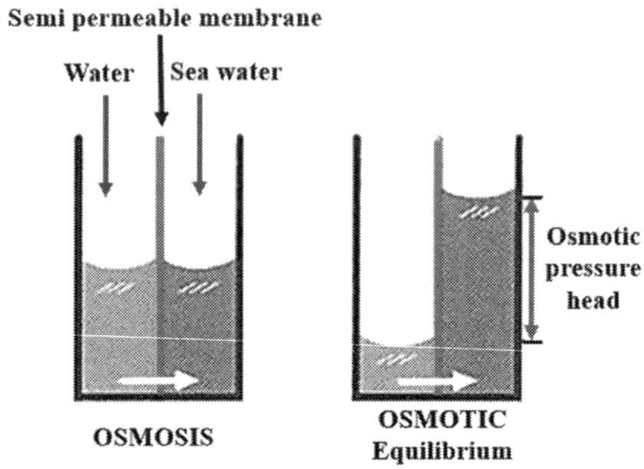

Figure 5.12 Osmosis process (Khulbe & Matsuura, 2018).

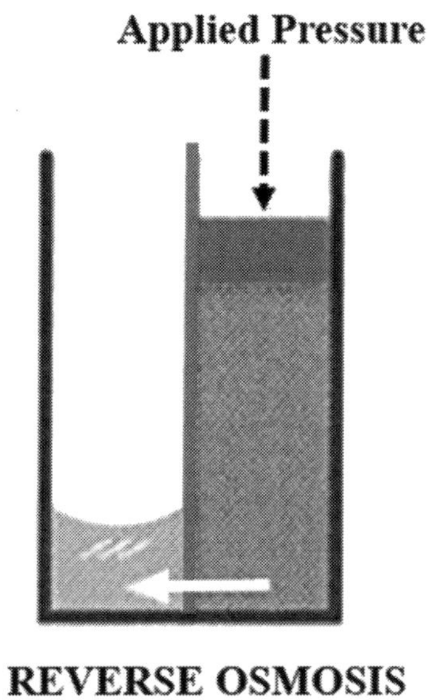

Figure 5.13 Reverse osmosis process (Khulbe & Matsuura, 2018).

harvested from the other side of the membrane as a low-pressure penetrating component. The pressure needed is related to the concentration of the salt solution on the refuse (concentrate) side of the membrane. It has the separation range of 0.0001−0.001 μm (i.e., 1−10 Å) particle size.

Pfeffer, Traube, and others studied osmotic phenomena with ceramic membranes as early as the 1850s. In 1931 the process was patented as a method of desalting water, and the term reverse osmosis was coined (Horvath, 1931). Modern interest dates from the work of Reid and Breton, who in 1959 showed that cellulose acetate films could perform this type of separation (Reid & Breton, 1959). Their films were 5–20 μm thick so fluxes were very low but, by pressurizing the feed salt solution to 1000 psi, they obtained salt removals of better than 98% in the permeate water. The breakthrough discovery that made reverse osmosis a practical process was the development of the Loeb–Sourirajan anisotropic cellulose acetate membrane (Baker, 2004). Anisotropic cellulose acetate membranes were the industry standard through the 1960s to the mid-1970s, until Cadotte, then at North Star Research, developed the interfacial polymerization method of producing composite membranes (Cadotte, 1985). Fluid Systems introduced the first commercial interfacial composite membrane in 1975. The construction of a large seawater desalination plant at Jiddah, Saudi Arabia using these membranes was a milestone in reverse osmosis development (Muirhead, Beardsley, & Aboudiwan, 1982).

Reverse osmosis membranes can be divided into three main categories as follows:

a. Membranes of seawater and saline water desalting used for solutions with concentrations of 0.5%–5% (based on weight percent) and operating pressure of 10–60 bars.

b. Membranes of low-pressure nanofiltration that are used for desalting of salt solutions with a concentration of 200–5000 ppm at pressures of 5–10 bars.

c. Membranes of hyperfiltration are applicable for the separation of solutes from solutions with an organic solvent.

The water flux J_i, is linked to the pressure and concentration gradients across the membrane by the equation:

$$J_i = A(\Delta p - \Delta \pi) \tag{5.2}$$

Where Δp is the pressure difference across the membrane, $\Delta \pi$ is the osmotic pressure differential across the membrane, and A is a constant. As this equation shows, at low applied pressure, when $\Delta p < \Delta \pi$, water flows from the dilute to the concentrated salt-solution side of the membrane by normal osmosis. When $\Delta p = \Delta \pi$, no flow occurs, and when the applied pressure is higher than the osmotic pressure$\Delta p > \Delta \pi$, water flows from the concentrated to the dilute salt-solution side of the membrane (Baker, 2004). Some basic information on reverse osmosis process is presented in Table 5.7 (Nath, 2017).

Table 5.7 Characteristics of the reverse osmosis process.

Characteristic	**Range/description**
Driving force	Chemical potential gradient
Transport mechanism	Diffusion
Separation principle	Solution-diffusion, preferential sorption
Size of retained species	1-10 Å
Membrane structure	Asymmetric, thin-film composite
Pore size	<2 nm
Types of reverse osmosis ilters	Desalination system, brackish water system, water treatment plants

From: Nath, K. 2017. Membrane separation processes (PHI Learning Pvt. Ltd.).

The flux of the diffused water and the amount of salt excretion are key parameters in the performance of reverse osmosis processes and nanofilters. Under certain conditions, the flux of the diffused water and the amount of salt excretion is one of the inherent properties of membrane performance. Basically, these two properties are influenced by the following factors: Pressure: by increasing the effective feed pressure, the flux of the diffused water and preferential sorption of water through the membrane increases (Fig. 5.14).

Temperature: If the temperature rises and the other parameters are constant, water flux increase and the flow of salt will increase since viscosity decrease (Fig. 5.15).

Solute concentration: Flux declines with increasing solute concentration because of the increase in osmotic pressure difference (Fig. 5.16).

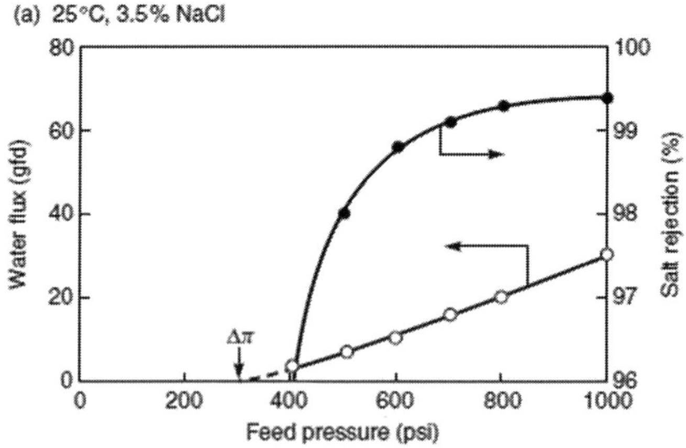

Figure 5.14 Effect of pressure on the properties of the membrane (Baker, 2004).

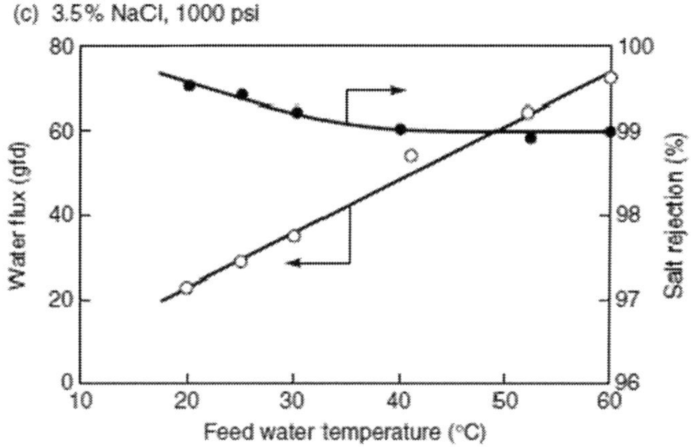

Figure 5.15 Effect of temperature on the properties of the membrane (Baker, 2004).

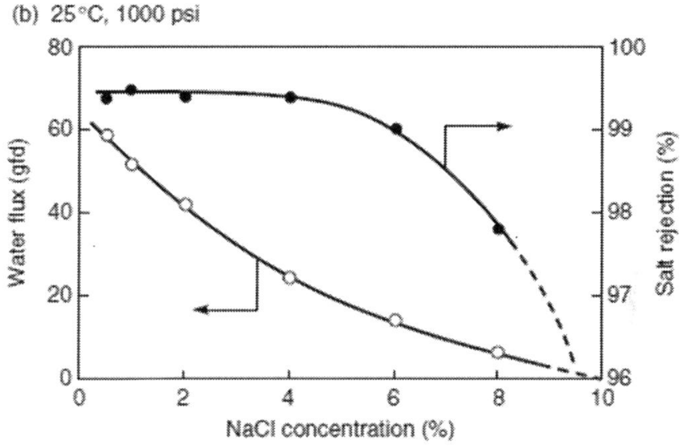

Figure 5.16 Effect of solute concentration on the properties of the membrane (Baker, 2004).

Recovery factor: Recovery is the ratio of pure water flow to feedwater flow. The consideration of recovery factor actually represents plant capacity and is generally in the range of 75%–95% with 80% being the practical maximum. At high recovery factors, there is a greater salt concentration in the process water as well as in the brine. At higher concentrations, salt precipitation on the membrane increases causing a reduction in separation efficiency.

Reverse osmosis is most generally recognized for its use in drinking water filtration from seawater, eliminate the salt and other effluent materials from the water. Other uses of RO include the removal of organics, salts, and silica ahead of deionizers in the boiler feedwater, removal of inorganic salts, phosphorus and nitrogen compound in the municipal wastewater treatment and also the demineralization of seawater and brackish water in the production of potable water. Several important applications of reverse osmosis are desalination of brackish water, natural color concentrates, the concentration of landfill leachate, the concentration of fruit juice, and concentration of digested sludge liquid. The following is another application of reverse osmosis, which is the treatment of dairy wastewater to value water and milk solids. Reverse osmosis treatment has been used to recover milk solids from dairy washing water which is an effective environmental strategy (Barão, Vieira Salla, Miorando, Hemkemeier, & Cadore Favaretto, 2019). The water from this process can be used as an on-site coolant as well as produce a milk-based white stream that is added to fermented milk drinks (Brião et al., 2019). The reverse osmosis membrane used has a 3.8 inch spiral-wound shape and an effective surface area of 7.1 m^2. Fig. 5.17 shows the outline of the process performed.

Concentrated dairy wash water using RO membrane, recycled condensed milk solids have the same composition as dry milk. Retentate was pasteurized at 75°C for 1 min and then frozen at −18°C to produce fermented milk and dulce de leche (Brião et al., 2019). The obtained permeate was evaluated according to various rules and instructions, and the search for

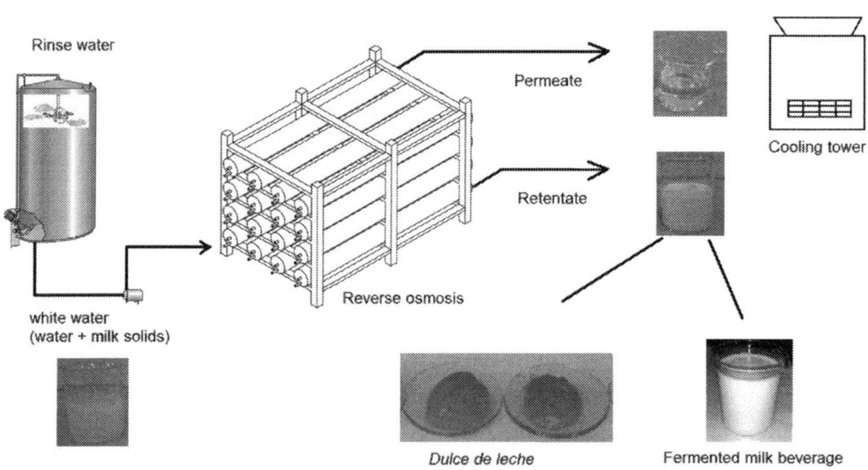

Figure 5.17 Diagram of wastewater treatment process of dairy industry (Brião et al., 2019).

end-use was carried out taking into account the health of consumers and the performance of equipment such as boilers and cooling towers. The potential profit could be as much as US $ 349,000 per year for the production of 1000 m^2 of dairy products per day (Brião et al., 2019).

5.2.3 Electrodialysis

The electrodialysis process is a process in which charged membranes are used. In this case, the purpose of the process is to remove the ions from aqueous solutions. In the 1960s, the only electrodialysis was used in the industry and ED, developed in the early sixties by Ionics of Watertown in the USA. The large-scale industrial utilization of electrodialysis began about 20 years ago, the principle of the process has been known for about 100 years. The development of electro-membrane processes began in 1890 with the work of Ostwald, who studied the properties of semipermeable membranes and discovered that a membrane is impermeable for any electrolyte if it is impermeable either for its cation or its anion (Bernardes, Rodrigues, & Ferreira, 2016). The first commercial equipment based on ED technology was used in the 1950s as an efficient and economical way to desalination of the brackish water (Strathmann, 2004).

In other words, electrodialysis is a process in which ions are carried by an ion-exchange membrane from a feed solution to another under the influence of a potential electrical difference. During this transition, anions are able to pass through the anion-selective membranes but are blocked by cationic membranes, conversely, this process occurs for cations (Nath, 2017). Electric charges allow ions to pass through an ion-exchange polymer membrane. The principle of electrodialysis is painted in Fig. 5.18 (Strathmann, 2004). Which shows a basic diagram of a typical order of ED cells, which include a series of cationic and anionic exchange membranes arranged in the alternating pattern between a cathode and anode which are used to form individual cells. The cell contains a volume of two adjacent membranes. The real electrodialysis unit, which is called a stack, may have several hundred of membranes (Bernardes et al., 2016).

The driving force in the electrodialysis process is the electrical potential between the anode and the cathode. The electric field applied causes the migration of positive ions (cations) to the cathode and the negative ions (anions) to the anode. During migration, cations pass through the cation

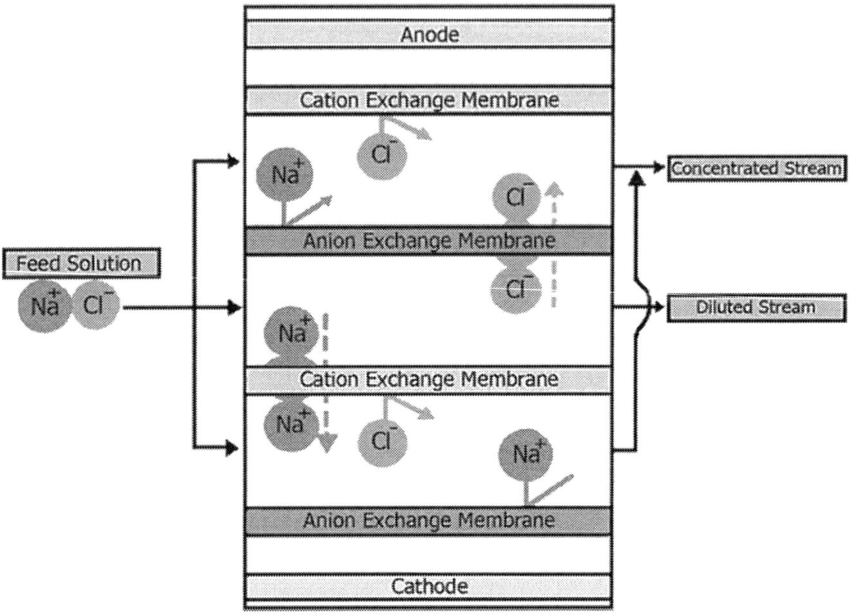

Figure 5.18 Principle of the desalination process by electrodialysis (Bernardes et al., 2016).

exchange membranes, but by anion exchange membranes are maintained; the same effect occurs for anions (Bernardes et al., 2016). Some basic information on electrodialysis are given in Table 5.8:

The process of electrodialysis is used in the following cases: electrodialysis in water cure, electrodialysis cure of refinery wastewater, electrodialysis cure

Table 5.8 Characteristics of the electrodialysis process.

Characteristic	Range/description
Driving force	Electrical potential gradient
Transport mechanism	Counter ions via ion exchange membranes, coupled transport
Separation principle	Donnan exclusion mechanism
Size of retained species	Coions, macroions
Type of membrane	Ion exchange and bipolar, semipermeable
Pore size	Nonporous
Main applications	Desalination of salt water Whey demineralization Pickling bath recycling Pharmaceutical applications Stabilization of wine

Nath, K. 2017. Membrane separation processes (PHI Learning Pvt. Ltd.).

of tannage wastewater, electrodialysis cure of phosphate solutions, electrodialysis cure of metal- cyanide collection. Another group of processes is known as developing industrial membrane separation processes. Although these processes have been industrialized, their technology continues to expand. Gas separation and pervaporation are of these processes.

5.2.4 Gas separation

Gas separation is widely used in the world to separate one desired component from a gas mixture. The most common uses of membranes in gas separation include the separation of hydrogen from nitrogen, argon, and methane in ammonia plants. Membranes can also be used to produce nitrogen from the air. Furthermore, membranes can separate carbon dioxide from methane in natural gas operations. Gas separation over the past 20 years has become an important membrane industrial process, but the gas separation has a long history (Baker, 2004). A systematic study began with Thomas Graham who, measuring the amount of gas penetration through each diaphragm available to him for a period of 20 years (Graham, 1866). Fig. 5.19 provides a summary of the development of gas separation technology.

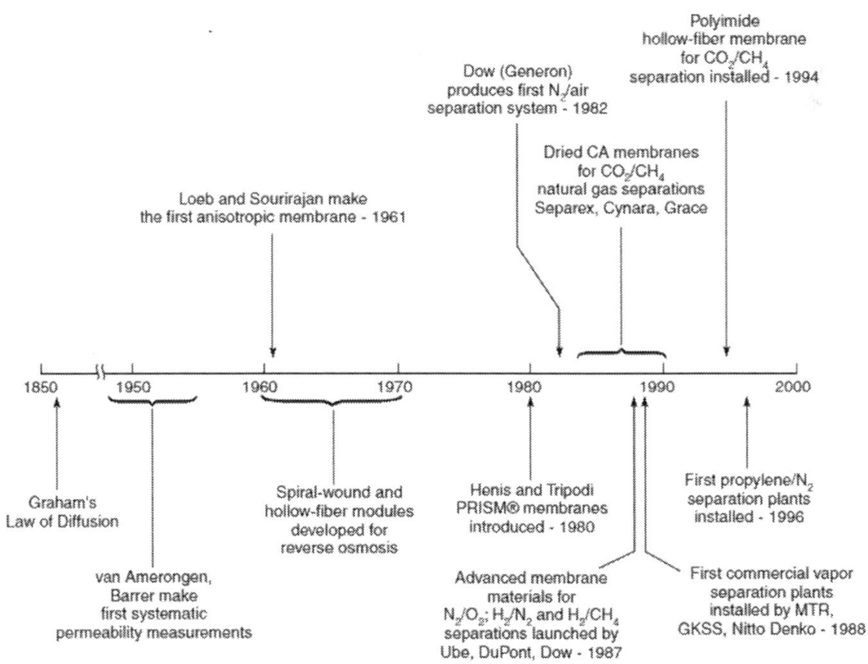

Figure 5.19 Milestones in the development of gas separation (Baker, 2004).

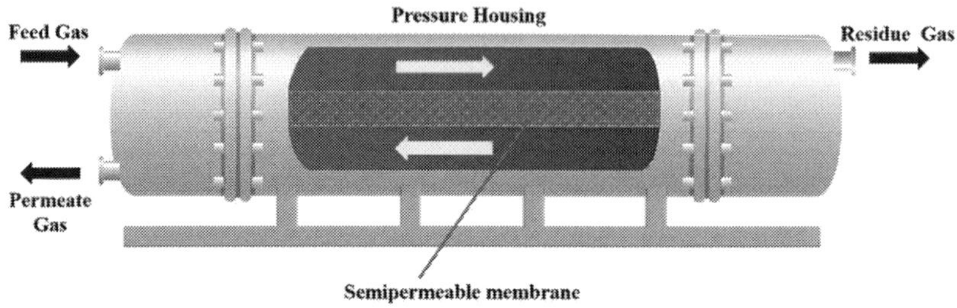

Figure 5.20 Schematic diagram of membrane gas separation. *From: Nath, K. 2017. Membrane separation processes (PHI Learning Pvt. Ltd.).*

Fig. 5.20 shows the common membrane process for gas separation. The feed is fed to one side of the membrane at high pressure. The opposite side has less pressure. The pressure difference in the membrane is the driving force for gas penetration in the membrane. Separation is achieved because of the differences in the relative transport rates of the feed components. Rapidly penetrating components are enriched during low-pressure penetration, while the slower components are concentrated in the remaining stream.

Three general transport mechanisms are commonly used to describe gas separations using membranes. They are Knudsen diffusion, molecular sieving, and solution-diffusion. Porous and dense membranes can be used as selective barriers to gas separation. Fig. 5.21 shows mechanisms for permeation of gases through porous and dense gas separation membranes.

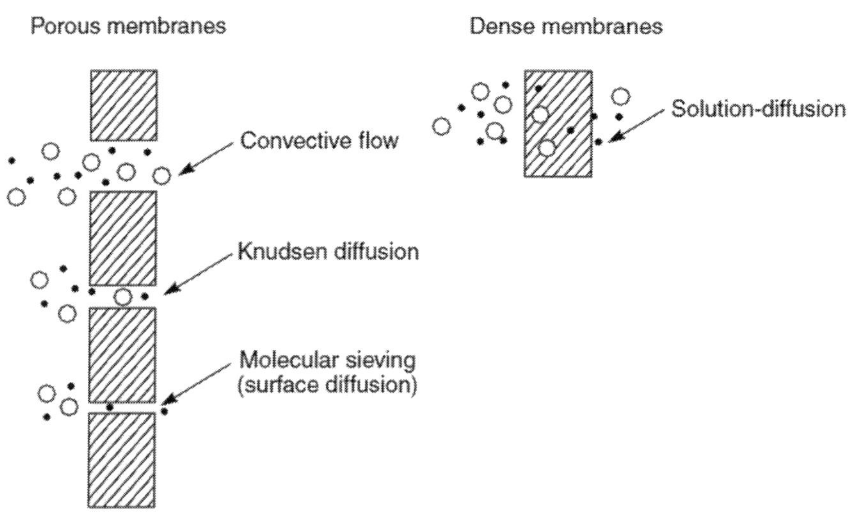

Figure 5.21 Mechanisms for permeation of gases through porous and dense gas separation membranes (Baker, 2004).

If the pores are relatively large—from 0.1 to 10 μm—gases permeate the membrane by convective flow, and no separation occurs. If the pores are smaller than 0.1 μm, then the pore diameter is the same size as or smaller than the mean free path of the gas molecules. Diffusion through such pores is governed by Knudsen diffusion, and the transport rate of any gas is inversely proportional to the square root of its molecular weight. The Knudsen diffusivity is calculated from the following equation:

$$D_{KA} = \frac{2}{3} r v_A \tag{5.3}$$

where D_{KA} is diffusivity in m^2/s, r is average pore radius in m and v_A is the average molecular velocity for component A in m/s.

If the membrane pores are extremely small, of the order 5—20 Å, then gases are separated by molecular sieving.

Solution-diffusion separation is relying on both solubility and dynamism factors. Gas permeation can be seen as a three-stage process in the solution-diffusion model:

- Adsorption and dissolution of gas at the polymer membrane interface.
- Gas diffusion inside and via bulk polymer.
- Desorption of gas into the external phase.

Of these three stages, the diffusion stage is the slowest, so this stage is the controller of the transmission speed.

Permeability coefficient P_A is the permeation of a pure gas A through a membrane of thickness l_m at steady-state, generally defined as:

$$P_A = \frac{N_A}{(P_1 - P_2)/l_m} = \frac{N_A}{\Delta P/l_m} \tag{5.4}$$

Where N_A is the normalized flux, P_1 and $P_2$are the upstream and downstream pressures respectively (partial pressures of penetrant A at the two sides of the membrane), and ΔP is $P_1 - P_2$.

For a two-component gas mixture that permeates through a polymer membrane, the choice of a polymer membrane over two different permeable gases, A and B, is usually expressed as the ideal perm-selectivity, α_{AB}. When the downstream pressure is negligible relative to the upstream pressure, α_{AB} can be written as the ratio of permeabilities:

$$\alpha_{AB} = \frac{P_A}{P_B} \tag{5.5}$$

The permeability coefficient, P_A can also be defined as:

$$P_A = D \times S \tag{5.6}$$

where, S is the solubility coefficient and D is the diffusion coefficient.

Expanding the permeability into diffusivity and solubility terms, the ideal selectivity can be expressed by:

$$\alpha_{AB} = \left(\frac{D_A}{D_B}\right)\left(\frac{S_A}{S_B}\right) \tag{5.7}$$

High permeability and high selectivity are the most important criteria in evaluating a membrane.

Membrane gas separation modules operate on the basis of selective permeation. As the gases adsorb on the surface, dissolve and penetrate into the polymer material, this technology uses this fact. Some of the most important applications of gas separation are expressed below.

Recovery of hydrogen from tail gases: highly pure hydrogen produced by palladium membranes is used in semiconductor industries. Six moles of hydrogen can be absorbed on 1 mole of palladium. Hydrogen molecules break down into atoms and are absorbed in the high-pressure side of the membrane. The electrons around the proton are then transferred to the palladium free electron band. The naked proton passes through the membrane at high pressure to the low pressure of the membrane and receives an electron at the low-pressure side of the membrane and forms a hydrogen atom. Eventually, two hydrogen atoms on the low-pressure side of the membrane form a hydrogen molecule (Nath, 2017).

Other applications of gas separation are removal of carbon dioxide, separation of oxygen and nitrogen, helium recovery from natural gas, and dehydration of air.

5.2.5 Pervaporation

Pervaporation is a partly new process that involves reverse osmosis and the separation of gas processes in a series of subscriptions. Pervaporation is a mixture of membrane permeation and vaporization. In the pervaporation membrane separation process for the selective penetration of one or more molecules from the feed mixture, a dense polymer membrane used (Nath, 2017). In the Pervaporation process, a fluid mixture is placed in contact with one side of the membrane, and the penetrating component is disposed on the other side of the membrane in the form of a vapor that is shown schematically in Fig. 5.22. To create the required driving force in this process,

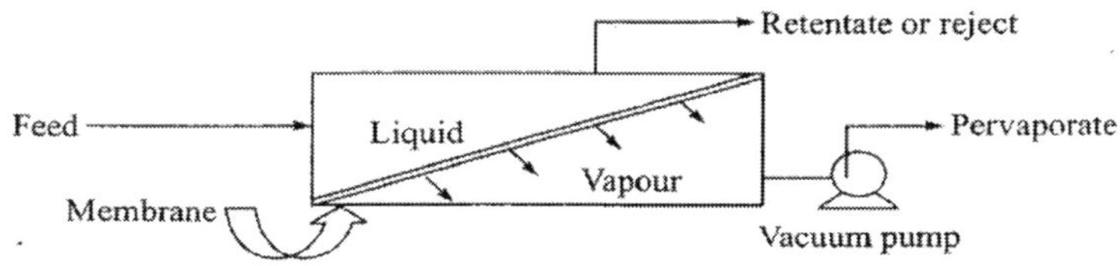

Figure 5.22 The schematic of pervaporation unit (Bai et al., 2015).

the vapor is cooled in the penetrating component side of the membrane, then condensate and the vapor pressure is reduced. So the driving force is to reduce the vapor pressure (Baker, 2004).

The origins of pervaporation can be traced to the nineteenth century, but the word itself was coined by Kober in 1917 (Kober, 1917). The process was first studied in a systematic fashion by Binning and coworkers at American Oil in the 1950s (Baker, 2004). The process was picked up in the 1970s at Monsanto by Eli Perry and others. More than a dozen patents assigned to Monsanto issued from 1973 to 1980 cover a wide variety of pervaporation applications (Strazik & Perry, 1973), but none of this work led to a commercial process. Academic research on pervaporation was also carried out by Aptel, Néel, and others at the University of Toulouse (Aptel, Challard, Cuny, & Neel, 1976; Neel, Nguyen, Clement, & Blanc, 1983).

The principles of pervaporation can be better understood by explaining the two stages of evaporation and membrane transport. In the evaporation process, the fluid temperature of the feed increases to saturated vapor, and when the feed enters the device, the saturated vapor contacts the membrane. In the membrane transport process, vapor from the feed side penetrates across the membrane to the permeate side.

Some of the advantages of this method expressed below:

- This method is useful for separating materials that are sensitive to heat and cannot be used distillation.
- This method is useful for the separation of the mixture with a boiling point near each other and the azeotropic mixtures in which the distillation method is not possible.
- With this method, the hybrid system can significantly save energy.
- This method is not contaminated.
- This method does not use any absorbent that needs to be regenerated.

The membrane selection for this process depends on the feed solution. Hydrophilic membranes and organophilic membranes are used to remove water from organic solutions and eliminate an organic species from gaseous effluents or aqueous respectively. In this method, composite membranes are also used, which consists of two layers. The first layer is porous, polymeric support and the second layer or the top layer is an active (perm selective) layer that is designed to absorb more of the chemicals desired.

Some of the most important applications of pervaporation expressed below:

Solvent Dehydration, control of ethanol concentration in the fermenter, recovery of methanol from methanol/MTBE/C4 mixture, separation of ethylene glycol-water mixtures, removal of VOCs from water.

A number of other membrane processes are in the category of to-be-developed technologies. The carrier facilitated transport is the most important of this category.

5.2.6 Facilitated transport membranes

In this category of membrane processes, liquid membranes are used as a carrier agent. The carrier adsorbs and reacts with a component of the mixture at the feed portion of the membrane and then penetrates the membrane and release the penetrating component at the product side of the membrane. The reformed carrier agent then returns to the feed portion of the membrane. Therefore, the carrier acts as a shuttle that transmits a component of the feed from the feed side of the membrane to another side of the membrane. The following figure shows the important times of progress and development of this process.

As shown in Fig. 5.23 in 1960, Scholander first demonstrates facilitated transport of oxygen with hemoglobin solutions. Great work was done by Ward et al. and Steigelmann and Hughes to develop this process and membrane in the late 1960s and early 1980s (Baker, 2004).

There are other processes such as dialysis that is not used in industries but are widely used in medical fields to remove blood toxic metabolites in patients with renal insufficiency. Synthetic kidneys have been developed based on the technology of dialysis membranes that reproduce cellulose. Inspired by the idea of artificial kidneys, the production of similar devices that could remove carbon dioxide and oxygen from blood was considered. This equipment is known as synthetic lungs and is used in the surgical treatment of patients with pulmonary insufficiency. One of the most important medical uses of membranes is drug delivery systems.

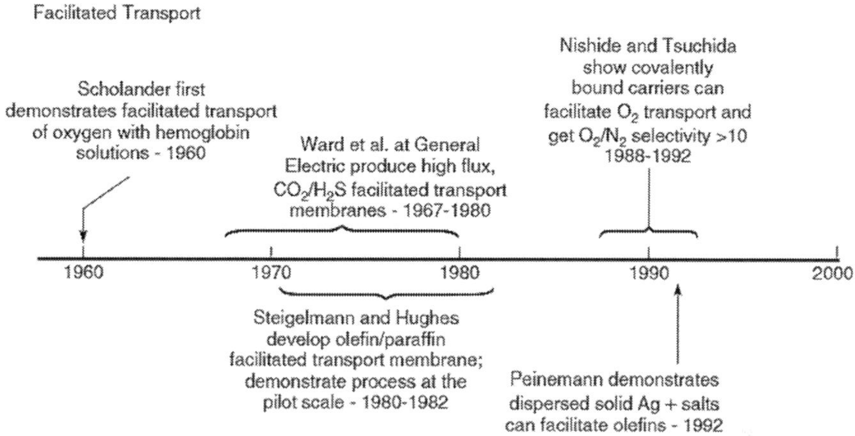

Figure 5.23 Milestones in the development of carrier facilitated transport (Baker, 2004).

5.3 Ion-exchange membranes

In the Ion-exchange membrane processes, membranes are used to carry an electric charge. This type of membrane can separate the charged particles selectively from solutions that contain other neutral components. The most important applications of the ion-exchange membrane process in the industry are electrodialysis and membrane alkaline cell processes (Baker, 2004). The first process in which the ion-exchange membrane technology was used at the industrial scale was electrodialysis. Electrodialysis was initially used to desalinate water. In some cases, the ion-exchange membrane process has been used to concentrate the seawater to obtain salt (Baker, 2004). In the early 1890s, experiments with an ion-exchange membrane were staged by Stewart (Baker, 2004). A few years later, Donnan developed the concept of membrane potential and led to the phenomenon of Donnan exclusion (Donnan. 1995). In Fig. 5.24, the times when the development of ion-exchange membranes has reached a turning point is shown (Baker, 2004).

The favorable characteristics of ion-exchange membranes are:

- High permanent selectivity for oppositely charged ions.
- Low electrical resistance.
- It is mechanically strong.
- High figure stability.
- After applying the thermal stress should not be wrinkled and deformed (Nath, 2017).

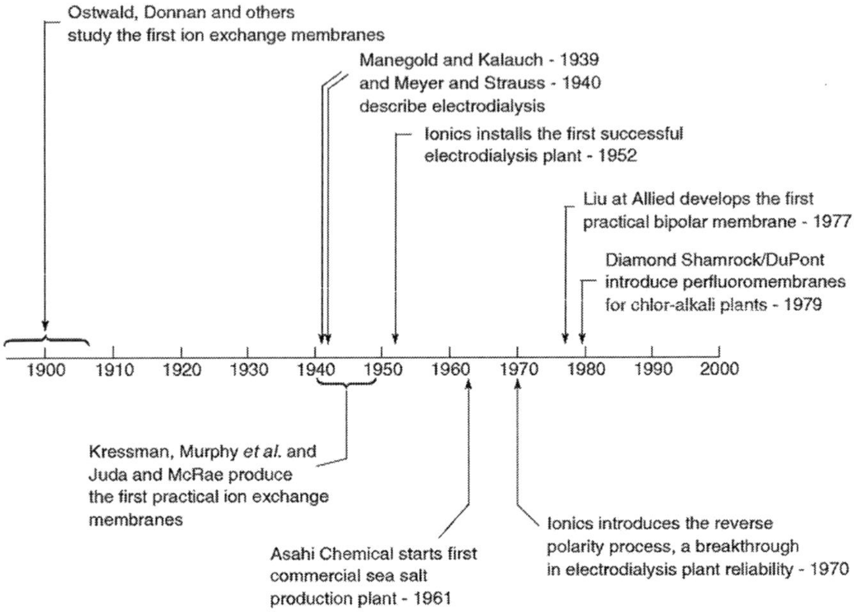

Figure 5.24 Milestones in the development of ion-exchange membrane processes (Baker, 2004).

In the process of the ion-exchange membrane, the polymeric surface of the membrane is completely covered by charged particles. This charged surface of the membrane prevents the passage of ions having the same charged from the membrane. For example, a positive-charged side of the membrane does not allow positive-charged particles to pass, and in return, negative-charged particles easily pass through the membrane, and so on, a negative-charged side of the membrane does not allow negative-charged particles to pass, and in return, positive-charged particles easily pass through the membrane. Fig. 5.25 shows the schematic of the anion and cation exchange membranes.

In the past, the main use of electrodialysis was in the desalination of saline water, but with the development of reverse osmosis, electrodialysis was less widely considered. Extending a combined process of electrodialysis and ion-exchange for desalination in cases where there is a need for very pure water so-called electrodeionization, the use of the ion-exchange method has thrived again, so today the most important application of electrodialysis is an electrodeionization process. Other applications of electrodialysis are

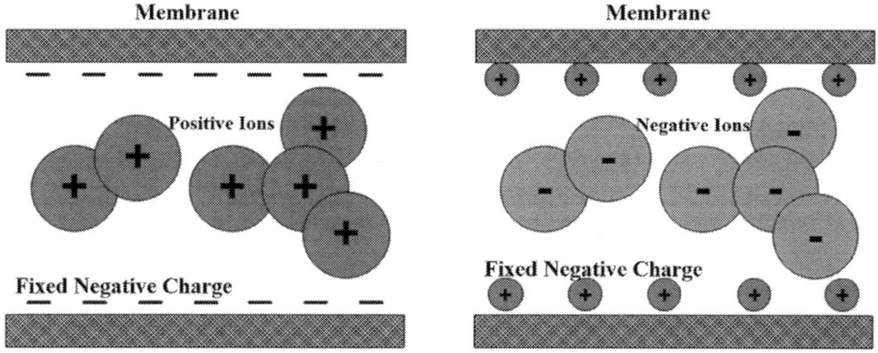

Figure 5.25 Cation and anion exchange membranes. *From: Nath, K. 2017. Membrane separation processes (PHI Learning Pvt. Ltd.).*

used to eliminate ionic impurities in industrial wastewater, to get water hardness and to desalinate in the food industries, e.g., in the milk producer industry.

The driving force in separation processes using ionic exchange membrane, such as electrodialysis, is the concentration gradient and potential gradient of electrical. However, it is common for ion-exchange membrane processes to consider the amount of charge transferred than the amount of material transferred since in this process there are two types of ions, anions, and cations, which transport in the opposite direction under electrical potential gradient. Due to the different sizes of icons, the number of different ions transferred in the ion-exchange membrane process can be in a very wide range. In the ion-exchange process, ions that have the same charge as the membrane plates are removed and only cause a very small current through the membrane. Therefore, in these membranes, few ions that are similar to the ions removed that the membrane can pass through itself. The amount of this transfer is 0%—5% of the total ions that the membrane can pass through itself, in other words, the total current that can be passed through the membrane. Other ions that have an ion charge opposite to the membrane plate can easily pass through the membrane, and almost form all of the current passes through the side of the membrane. The amount of this transfer is 95%—100% of the total ions that the membrane can pass through itself.

Ion-exchange membranes are divided into two large categories:

- Homogeneous membranes
- Heterogeneous membranes

In homogeneous membranes, the ionic charge is distributed uniformly on the surface of the membrane, in other words, in other words, the distribution of charged groups through the membrane is uniform. Homogeneous membranes when exposed to water are swelled by absorbing water uniformly, and the density of the membrane controls the extent of swelling in them. Below are a few ways to prepare them (Nath, 2017):

- Linking the polymerization of anionic or cationic fragments on the molecular chains of a premade film.
- Polymerization of mixtures of reactants that can tolerate condensation polymerization.
- Introduction of anionic or cationic parts into a polymer chain such as polysulphone and then by the dissolution of the polymer and casting into a film.

In heterogeneous membranes, ion-exchange groups are located on small sites that are distributed throughout the inert holding support that the role of this support is to cause the mechanical strength. They can be prepared by the following methods (Nath, 2017):

- First, they disperse the ion-exchange media in a polymeric solution that forms the polymer and then, by film casting and solvent evaporation, form the desired membrane.
- Dry molding of mixtures of the ion-exchange and film manufacturing materials.

5.4 Gas sorption

The gas separation process by using membrane technology has been considered from 30 years ago and continues to grow and develop in the current century. In order to perform gas separation operations selectively, membranes made of porous or dense material can be used. For pores with size between 0.1 and 10 μm, which are considered as relatively large pores, the motion of the bulk gas occurs in the pores and therefore no gas separation is performed. Among the applications of gas separation using industrial-scale membranes, we can mention the separation of hydrogen from nitrogen in pure ammonia gas flow. Hydrogen is a gas with small molecules that are not condensable and its permeability is significantly much more than other types of gases.

Membrane-relying gas separation can recourse to any of a number of techniques used to separate gases, either to give many products or to clean a single product. A gas adapts to the shape of a dish in which it is retained and accumulate the same density inside the dish, even in the presence of gravity and irrespective of the amount of material in the dish. If not confined to a container, gases, also known as vapors, will disperse into space. The atoms or molecules of matter in a gaseous state move freely among each other, and are, in most instances, packed more loosely than the molecules of the same material in the solid or liquid case (Ismail, Khulbe, & Matsuura, 2015). The most important characteristics of gases include:

- High compressibility: External forces can compress gas and decrease its volume; removing the external force allows the gas volume to increase.
- Thermal expandability: When a gas sample is heated, its volume increases, and when it is cooled its volume decreases.
- Low viscosity: Gases flow much easier than liquids.
- Low density: Gas densities are on the order of grams per liter, whereas liquids and solids are grams per cubic cm—1000 times greater.
- Infinite miscibility: Gases mix in any proportion, such as in air, and in any combination.

Important variables that describe a gas are pressure (P), temperature (T), and volume (V).

In gas mixtures, the gas behavior depends on the number, not the identity, of gas molecules the ideal gas equation used to each gas particularly and to the mixture as a complete. Besides these properties, all molecules in gas mixtures behave exactly the same way. Membrane technology has received significant attention from various industrial sectors and academics in their research because the technology offers the most relevant means of reducing environmental problems and costs. The gas separation membrane field is highly competitive both between companies developing membrane technologies and with other gas separation technology developers. Membranes are employed in:

- Nitrogen or oxygen separation from the air;
- Hydrogen separation from gases such as nitrogen and methane;
- Hydrogen recovery from ammonia plant crop flow;
- Recovery of hydrogen in oil refinery processes;
- Separation of methane from the other components of biogas;
- Enrichment of air by oxygen for medical or metallurgical purposes;
- Ullage enriched by nitrogen in inert systems designed to prevent fuel tank explosion;
- Removal of water vapor from natural gas and other gases;
- Removal of CO_2 from natural gas;
- Removal of H_2S from natural gas;

- Removal of volatile organic liquids (VOL) from an air of exhaust streams.

Inorganic membranes are increasingly being a probe to separate gas mixtures. Such membranes usually have much higher gas fluxes as compared to polymeric membranes. Furthermore, inorganic membranes are well-known to be more thermally and chemically stable than polymeric membranes (Ismail et al., 2015).

5.5 Nanostructure membranes

Most membranes that are used on an industrial scale are made from polymer materials. However, the use of substances that are less common is nowadays expanding. The ceramic membrane is a special type of microporous membranes that are used in UF and MF processes because, in these processes, solvent resistance and thermal stability are important. Membranes made of dense metals such as palladium-based membranes have acceptable performance in the process of separating hydrogen from gas mixtures. When the purpose is selecting of membrane materials for a particular separation process, various parameters have to be considered, including lifetime, mechanical integrity, operating conditions, the efficiency of separation and Flux passing through the membrane. In membrane operations, the most important features of the membrane are selectivity, penetration rate, concentration polarity, thickening, and antifouling properties.

Nanocomposites are materials that have a solid structure in which the distance between the phases is leastwise formed of a dimension with nanoscale size and general form of an inorganic matrix set in the organic phase, or vice versa, from an organic matrix set in the inorganic phase. Nanocomposite materials can exhibit individual properties such as mechanical, chemical, electrical, optical and catalytic properties. Nanocomposite membranes provide such conditions for the permeation of the desired component while acting as a preventive curtain against other ill-favored components. The quality of the interface between nanoparticles and organic polymers plays an important role, which can be greatly improved by modifying the base polymeric matrix and mineral nanofillers. In membranes made of nanocomposites, glass or rubber polymers can be used in the structure of membrane as an organic base matrix. Many nanoparticles can be used for exerting in the structure of mineral nanofilters. Those that can be mentioned include metal oxides, pure metals, zeolites, carbon, and other mineral materials. Specifically, TiO_2, Fe_3O_4, AgO, MgO, SiO_2, and Al_2O_3 from metal oxides, nanosilver from pure metals, zeolite 4A, silicate and ZSM-5 from zeolites, and carbon nanotubes, C60 fullerenes from carbon materials can be named.

Nanocomposite membranes are very important in membrane-based processes, especially new and up-to-date applications. Nanocomposite membranes have much more advantages than other common membranes, which include:

- Nanocomposite membranes have more permeability than conventional membranes.
- The required pressure in systems containing nanocomposite membranes is less than other membrane processes.
- Nanocomposite membrane systems are structurally compact systems.
- The required energy in systems containing nanocomposite membranes is less than other membrane processes.
- The required membrane surface area in systems containing nanocomposite membranes is less than other membrane processes.
- Systems containing nanocomposite membranes have a smaller footprint for industrial applications than other membrane processes.
- Systems containing nanocomposite membranes have higher mechanical stability than other membrane processes.
- Systems containing nanocomposite membranes have higher thermal stability than other membrane processes.

Nanocomposite, like all other nanoscale materials, is very important in the processes of gas separation, liquid separation, fuel cell, and biomedical applications.

5.6 Catalyst coated membranes

The catalyst coated membrane is a polymeric membrane coated on its side with a layer of catalyst.

5.7 Polymeric membranes

Only hydrogen membrane technology based on polymer materials, widely used on a commercial scale. Monsanto introduced the first hydrogen membrane process in the early 1980s. Subsequently, polymer membranes have been used in hydrogen recycling in various applications, such as purification processes in the ammonia and beverage process, and for adjusting the hydrogen/carbon monoxide ratio in syngas plants (Doong, 2012). The environment of the permeating component is an important factor in determining the diffusion coefficient of that component, and this concept is essential for the study of polymeric membranes. Polymers can be categorized into two main groups, glass, and rubber. The reason for the soft and elasticity of a rubber polymer can be attributed to the free movement of polymer molecules due to the thermal motion and also the pivotal motion of the polymer backbone. Extremely high diffusion coefficients in rubber membranes result from this thermal motion in parts of the polymer.

Due to the existence of a space hindrance caused by the spatial deposition of the atoms along the polymer backbone, there is no free movement of parts of molecules in glass polymers. As a result, the glass polymer is stiff and hardy solid. From the foregoing, it can be concluded that the limitation of the thermal motion in glass polymers makes their diffusion coefficients to be very low. Increasing the thermal energy to overcome the limited movement of the steric hindrance in the backbone of the polymer can reach sufficient value if the temperature of the glass polymer is rising. The glass polymer at this temperature changes to a rubbery polymer, in which this particular temperature is specified as the glass transition temperature. In Fig. 5.2, the diffusion coefficient is shown in terms of molecular weight for water as a liquid, two rubbers (natural and silicone) and polystyrene as a glass polymer to make a quantitative comparison. The diffusion coefficients for small molecules such as hydrogen and helium in polymer membranes do not differ significantly and are in a comparable range, and eventually, the difference is about a factor of two or three. The collisions of these small molecules are only dealing with one or two immediate neighbors. For these small molecules, the local environment in all three polymers referred to in Fig. 5.26 is not completely different from water. However, for large molecules with a molecular weight between 200 and 300 and a molecular size between 6 and 10 Å, the local environment is completely different in nonidentical media. Different parts of the polymer chain in a polymer membrane are involved in the movement of the diffusive particles within the membrane. Since it is unlikely that the particle motion is facilitated by other particles, it can be concluded that the diffusion coefficients in the polymer membranes are very less than ones in diagram water. In addition, the movement of different polymer segments in flexible rubber membranes is very different from hard polymer membranes. More flexibility and easier rotation of the rubber polymer chains than the hard polymers are very significant. The basis of this flexibility difference between rubber polymers and hard polymers is due to the elastic properties and also the difference in diffusion coefficients.

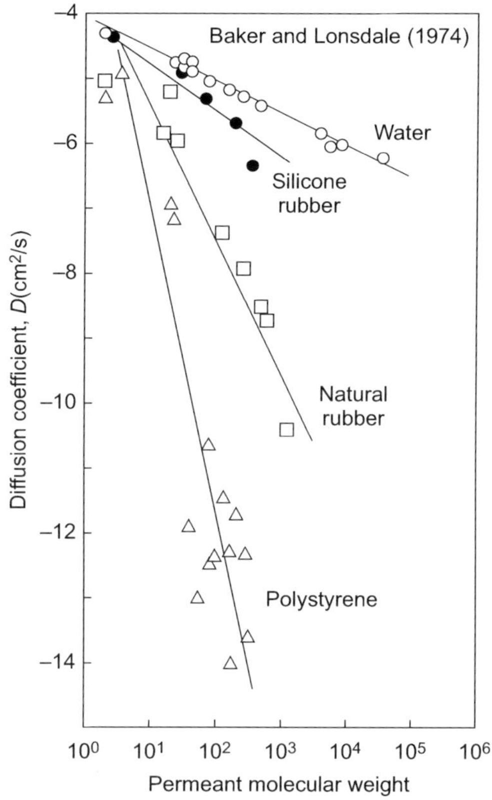

Figure 5.26 A quantitative comparison between the diffusion coefficient in terms of molecular weight for water as a liquid, two rubbers (natural and silicone), and polystyrene as a glass polymer.

In Fig. 5.27 the variation of the diffusion coefficients in terms of the Constituent material changing in the membrane matrix is well illustrated. In this case, the constituent material of the polymer matrix is ethyl cellulose and the gradually added material is dichloroethane. In this figure, the change in the diffusion coefficient is shown by the change in the polymer material of the membrane. To make the desired changes in the structure of the polymer matrix, the concentration of dichloroethane in the polymer matrix has been changed from less than 1% to more than 90%, i.e., from a very small amount to a very high amount.

The process of changing the structure of the polymer in terms of increasing the concentration of dichloroethane is such that at first, the

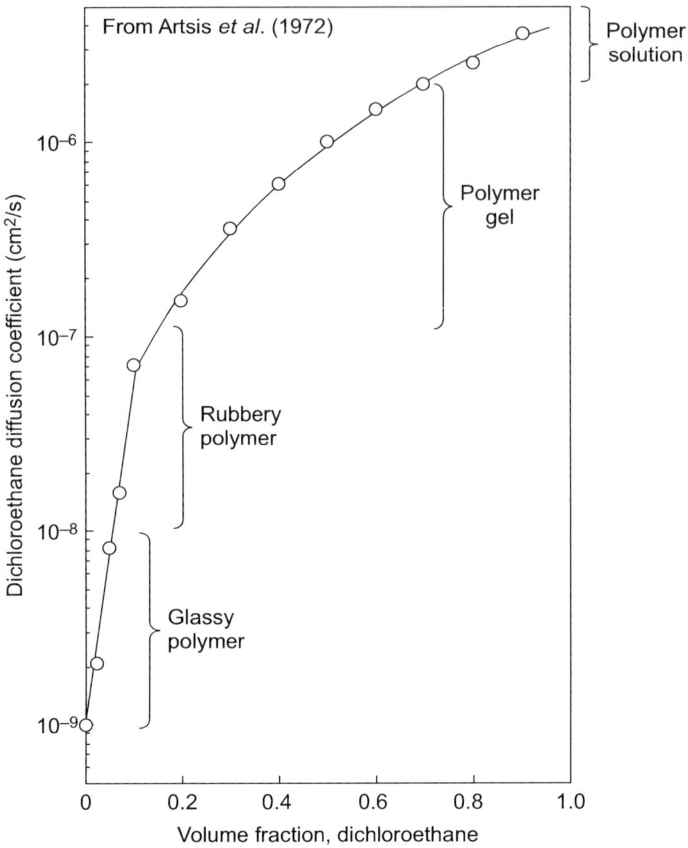

Figure 5.27 The change in the diffusion coefficient is shown by the change in the polymer material of the membrane.

polymer converts from glass to rubber, and then into gelatin state and with a very high concentration of dichloroethane to a dilute polymeric solution. The transition temperature of ethylcellulose, which is a glass polymer, is about 45−50°C. When the concentration of dichloroethane in a polymer is less than 5% vol., the polymer matrix is in the form of a glass polymer and the diffusion coefficient of dichloroethane is about $1 \times 10^{-9}\frac{cm^2}{s}$ to $5\times 10^{-9}\frac{cm^2}{s}$. The transition temperature of the glass polymer reaches below the temperature of the experiment when the dichloroethane is sufficiently dissolved in the polymer and its concentration in the polymer reaches up to 5% vol. Therefore, the polymer chains are sufficiently free to rotate and the hard polymer converts to a rubber and flexible polymer.

5.8 Ceramic membranes

Although ceramic membranes have been known for centuries, while polymers are new to the market, polymer membranes are dominant in the market. The best reason for this is money (usually, the cost of a ceramic membrane is three to five times the cost of polymer). Ceramic membrane materials are more expensive and their production process is complex and multi-stage, and their packing density is low in each membrane module. With all this, ceramic membranes have many advantages (Gitis & Rothenberg, 2016). In the previous sections, our focus was on membranes made of organic polymer materials; furthermore, polymer membranes are most widely used at the commercial scale. But, it should not be forgotten that the use of membranes made from less commonly used materials is currently expanding. In the processes of ultrafiltration and microfiltration, in which solvent resistance and thermal stability are important in them, the use of ceramic membranes that are classified as microporous membranes has been considered. Ceramic membranes have been manufactured by various companies for use in ultrafiltration and microfiltration processes. In the manufacture of ceramic membranes that are part of the microporous membranes, materials such as aluminum, titanium, and silicon oxides are used.

The main advantages of the ceramic membrane against polymer membrane are their long life and their high flux (Gitis & Rothenberg, 2016). The flux through ceramic membranes $\left(360\,\mathrm{L}/\left(\mathrm{m}^2\cdot\mathrm{h}\right)\right)$ is reported to be twice as high as the flux passing through the polymer membrane (Uhlhorn, Keizer, & Burggraaf, 1992). The superiority of ceramic membranes to polymeric membranes is that ceramic membranes are not chemically reactive (are inert substances) and stable at high temperatures, while polymer membranes are lack of such properties. Therefore, the use of ceramic membranes is very common in sensitive industries such as food, biotechnology, and medical industries, where membrane stability is very important in very high temperatures. In ultrafiltration and microfiltration operations, ceramic membranes are used with a diameter between 0.01 and 10 μm. In the following table, some of the advantages and disadvantages of ceramic membranes and their applications are given when benefits are important (Gitis & Rothenberg, 2016). Table 5.9 provides some of the advantages and disadvantages of ceramic membranes and their applications important (Gitis & Rothenberg, 2016).

Table 5.9 Some of the advantages and disadvantages of ceramic membranes and their applications.

Advantages	Disadvantages	Current status	Applications
Thermal stability (>200°C)	Low hydrothermal stability of composite membranes with a silica top layer. Complicated sealing at high temperatures	Surface modification to improve hydrothermal stability	Reactions at high temperature, pervaporation, gas separation
Resistance to organic solvents	Expensive source materials, complex processing; relatively high capital installation costs		Separation of organic liquids and oils, wastewater treatment
Chemical stability over a wide pH range	Difficult sealing and module construction		Chemical cleaning, separation at extreme pH, recovery of acids/bases
Long-time operational stability, no aging, potentially lower life cycle cost	Low packing density		Small-scale applications
Mechanical stability under large pressure gradients	Brittleness incurs special configurations and supporting systems		Operation at a high flux (up to 5001/(m2h)), backwashing at high flux
High structural integrity	Relatively high installation and modification costs in case of defects	Depends on the separation process	
Uniform pore size distribution	Difficulty to achieve high selectivity on a large scale		Precise size-based separations in pharma and bio applications
(Electro)catalytic and electrochemical activity easily realizable	Low permeability of high-selectivity (dense) membranes at medium temperatures		Solid oxide and other dense electrolyte fuel cells

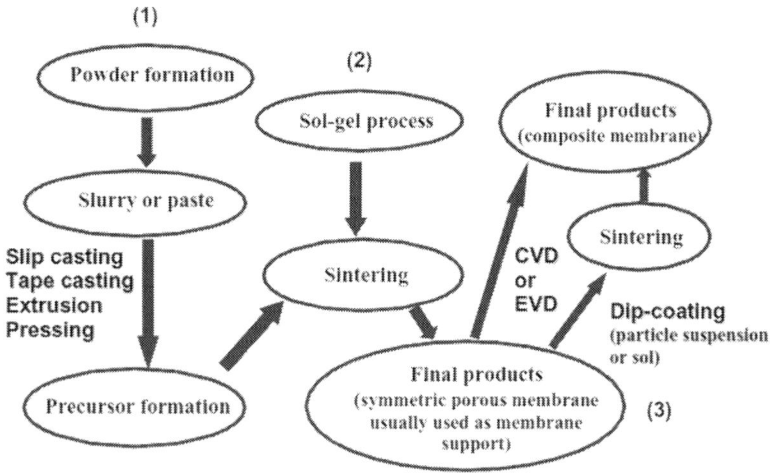

Figure 5.28 A generalized flow sheet for the preparation of ceramic membranes using various conventional methods (Uhlhorn et al., 1992).

Preparation of ceramic membranes has three steps: (1) establishment of particle suspensions; (2) impaction of the particles in the suspensions into a membrane pioneer with a certain form such as flat sheet or tube and (3) fixation of the membrane pioneer by a heat treatment at high temperatures (Uhlhorn et al., 1992). Fig. 5.28 shows a generalized method for preparation of symmetric and composite ceramic membranes that include different methods such as slip casting, tape casting, extrusion and pressing which turn the particle suspension (slurry or paste) into a membrane pioneer.

Fig. 5.28 shows that not important which process is selected in preparation of the ceramic membranes, the final membrane product can only be obtained through a firing step (Li, 2007).

Ceramic membranes can be manufactured from many materials. The strict definition of ceramics as inorganic nonmetallic material prescribes the use of metal oxides or zeolites, and actually, alumina and silica are among the most common membrane pioneer (Gitis & Rothenberg, 2016).

The following table (Table 5.10) is expressed a number of applications ceramic membrane.

5.9 Advanced oxidation processes (AOPs)—membrane hybrid systems

In the process of water purification for the separation of organic and inorganic impurities, there are a number of very advanced technologies

Table 5.10 Applications of ceramic membranes in liquid-phase separations (Gitis & Rothenberg, 2016).

Area	Membrane type	Application examples
Food and beverage industry	MF and UF	The concentration of milk, the concentration of protein, clarification of fruit juice, clarification of beer and wine, removal of microorganisms at fermentation
Biotechnology and pharmaceutical industry	MF and UF	Microorganism separation and cell debris filtration, plasma separation
Chemical and industrial applications	MF and UF	Oil-water separation, purification of used oil, removal of precipitated heavy metals and solids
Recovery and recycling	UF and NF	Drinking water and wastewater treatment

that can be used, such as chemical technologies, physical technologies, and some technologies that are a combination of these two methods. The basis of the method for advanced oxidation processes is that oxidizing radical particles are produced in situ, which have high reactivity with micropollutants.

Up to now, many studies have been conducted on the industrial application of advanced oxidation technologies and, in addition, newly discovered nature-friendly technologies have been developed on the basis of oxidation technologies. Advanced oxidation processes (AOPs) have significant characteristics that have led many companies to use these technologies in urban and industrial sewage treatment plants and to remove the pollutants of surface and underground water. One of the most important of these features is their unique and reliable ability to remove water pollution and cost-effective economic and sustainability of these methods.

The concept of hybrid membrane systems means that through the combination of membrane separation processes with conventional technology in the process industry, a new system can be achieved that has higher energy efficiency and less adverse environmental effects. One of the hybrid membranes is the organometal membranes. It is a polymer matrix that grains of zirconia are distributed on the surface of it. One of the important advantages of organometal membranes is that they have a very high-flux rate compared to old polymer membranes. Another process of the hybrid membrane is the electro deionization process that is used in semiconductor manufacturing, medical industries, and production of very pure water for power plants.

The distillation-pervaporation process is a hybrid process that is developed to obtain ethanol with a purity of 99.85% from mixtures of water and ethanol. In the process of separating alcohol from fermentation liquors, this technology has been used to remove alcohol, so that continuously increasing alcohol cannot prevent the production of microorganisms.

5.10 Photocatalytic membranes

By definition, the photocatalyst is a substance that can be produced by absorbing ultraviolet radiation, infrared radiation, visible light, and transubstantiation the chemical substances involved in the chemical reaction. After each interaction between the reactants and the photocatalyst, the chemical composition that existed before the reaction was recovered. In fact, photocatalytic membranes refer to those membranes that are made of combining a photocatalyst with a common membrane. The photocatalyst membrane is interesting because the photocatalytic reaction can take place on the membrane surface without the accumulation of nanostructured particles, the difficulty of reusing and losing photocatalyst particles (Chen, Hu, Xie, & Wang, 2018). The efficiency and effectiveness of a membrane as a separation barrier depends on a number of factors, including the type of membrane, the pore structure, and sizes, polarity, roughness, hydrophilicity as well as its mechanical properties. In other words, Materials in which one or more phases with nanoscale dimensions are embedded in another matrix phase are called nanocomposites. The motive behind fabricating such materials is to create synergies between the various phase constituents to result in smart materials capable of meeting or even exceeding the intended design expectations. Membranes have been functionalized through the incorporation of engineered nanoparticles deposited on their surfaces or embedded into their matrices to result in tailored properties such as the capability to bind specific contaminants or catalyze degradation reactions. The resulting mixed matrix membranes play a multifunctional role in water treatment. Variables such as the nanomaterial loading, the matrix material, degree of particle dispersion, size, shape, and orientation of the nanoscale phase as well as interactions between the phases have a bearing on the overall properties of the hybrid material. In photocatalytic membranes, nanoscale inorganic photocatalysts are embedded in a membrane matrix to enhance the properties of the resultant polymer. Photocatalytic membranes can be broadly classified into four categories, based on the location of the nanoparticles on the membrane:

- Conventional nanocomposites
- Thin-film nanocomposites

- The thin-film nanocomposite with nanocomposite substrate
- Surface-located nanocomposites

Materials for the production of photocatalytic membranes should have thermal and chemical stability in the application field and depend on these factors. Most of the semiconductor photocatalysts are inorganic materials such as TiO_2, ZnO_2, Bi_2O_3, etc. The studied membranes can often be produced from various materials, including polymers, metals, ceramics and even zeolites (Uhlhorn et al., 1992). A number of polymers have successfully been used as supports for photocatalysts, and these include polymers such as polyamide, polyvinylidene fluoride (PVDF), polysulfone (PSF), polyethersulfone (PES), sulfonated polyethersulfone (SPES), polyurethane (PU), polyethylene terephthalate (PET), polyester, polyacrylonitrile (PAN), and polytetrafluoroethylene (PTFE). Some of these polymers are listed in Table 5.11.

The photocatalyst nanoparticles can either be sedimented onto the membrane surface or dispersed in the polymer dope solution previous to membrane foundry. Several other methods have been proposed and tried for the fabrication of polymeric and ceramic photocatalytic membranes. The overall objective is to incorporate the photocatalyst onto the membrane to impart some photocatalytic activity to the membrane (Kuvarega & Mamba, 2016). The following table (Table 5.12) shows a summary of photocatalytic membranes with different substrates and their applications.

An important step in the preparation of a photocatalytic membrane is to connect photocatalyst to various supports. Various fabrication methods have been investigated, such as dip coating, spin coating, electrospinning, physical deposition (vacuum filtration), and immersion precipitation.

Dip Coating: an easy method that is widely used to prove photocatalyst on both polymer membranes and inorganic membranes.

Table 5.11 Summary of polymers used for membranes fabrication (Chen et al., 2018).

Name	Abbreviation	Formula
Cellulose acetate	CA	$(C_6H_7O_2(OH)_3)_n$
Polysulfone	PS	$(C_{27}H_{26}O_6S)$
Polyamide	PA	$(CONH_2)_n$
Polyvinylidene fluoride	PVDF	$(CH_2CF_2)_n$
Polyacrylonitrile	PAN	$(C_6H_4 - SO_2 - C_6H_4 - O)_n$
Polyethersulfone	PES	$(C_6H_9NO)_n$
Polyvinylpyrrolidone	PVP	$(C_2H_6O_2)_n$
Polyvinyl alcohol	PVA	

Table 5.12 Summary of photocatalytic membranes with different substrates and their applications (Chen et al., 2018).

Type of membrane	Application	References
TiO_2/Al_2O_3 Composite membrane	Methylene blue (MB) degradation	Choi, Stathatos, and Dionysiou (2006)
$FeTiO_2$/PS hybrid photocatalytic membrane	Removal of NO	Li et al. (2016)
$CoFe_2O_4$/rGO/PVDF/ carbon paper membrane	Removal of tetracycline hydrochloride antibiotics	Liu, Cheng, Zhao, and Zhu (2017)
TiO_2 hollow fiber membrane	Acid Orange 7 (AO7) degradation	Zhang, Wang, Schmeda Lopez, and Costa (2014)
TiO_2/carbon/Al_2O_3 membrane	Removal of natural organic matter	Wang, Chen, Yu, and Xie (2015)
CNT/ZnO/TiO_2 membrane	Removal of AO7	Bai et al. (2015)
g-C_3N_4/rGO/CA membrane	Removal of rhodamine B (RAB)	Zhao, Chen, Xie Yu, and Zhao (2016)
Silica/titania nanorods/nanotubes membrane	Removal of sodium dodecylbenzene sulfonate	Zhang, Xie, Chen, Zhao, and Zhao (2006)
TiO_2/carbon paper membrane	Removal of MB	He et al. (2016)

Spin Coating: In general, this technique is used to prepare thin membranes by spinning out the precursor solution over the edge of the support. The process is to add the coating solution onto the surface of the support, and the centrifugal force drives the solution to spread in a thin film where the solvent rapidly evaporates to produce a uniform coating layer on the substrate.

Electrospraying and electrospinning: The conventional electrohydrodynamic processes, including electrospraying and electrospinning, are easy, cost-effective, and flexible methods for fiber production. These methods are thereby carried out to produce fibrous membranes. These methods are used to fabricate highly porous membranes by electrospinning or electrospraying organic polymer nanofibers. These processes are driven by electrical forces.

Vacuum Filtration: in this method, the photocatalytic particles are physically placed on the surface of the membrane.

Immersion Precipitation: phase inversion via immersion precipitation is a widely used membrane preparation approach. A polymer plus polymer solution is cast on proper support and then submerged in a coagulation bath containing nonsolvent. Owing to the solvent and nonsolvent exchange, precipitation takes place (Chen et al., 2018).

The photocatalytic membranes also have unique properties including superhydrophilicity, antifouling performance, photocatalytic oxidation and reduction, separation, and antimicrobial ability. Even though photocatalytic membranes have shown great potential in water treatment, antifouling, disinfection, and energy conversion, many technical problems still need to be addressed before this technique can be practically applied in the industry (Chen et al., 2018).

5.11 Water and wastewater treatments plants

The increasing population and industry have led to the growing use of rivers, lakes, and reservoirs. The use of water, treatment, and reuse is a global challenge. Water and wastewater treatment processes are selected based on the initial water quality and according to the application and use of the water, for example, the quality of water needed for the industry varies with drinking water, so the method of purifying and treating water is different for both Membrane separation processes used for treatment of water and wastewater make it easy to reuse these sources, these processes have different features and applications. All current technologies produce a concentrated stream that should be discharged or treated in the environment (Bernardes et al., 2016).

Understanding membrane application requires an understanding of the characteristic of drinking water membrane process. Reverse osmosis, nanofiltration, electrodialysis reversal, ultrafiltration, and microfiltration are the membrane processes, which have an application to drinking water. The composition of membrane processes with other processes has become known as united membrane systems. Although a conventional nanofiltration process consists of a pretreatment and posttreatment process before and after the nanofiltration process, which could be described as integrated, this is described as conventional. The coupling of a microfiltration and a nanofiltration or coagulation, sedimentation, and filtration with a nanofiltration are accepted examples of integrated management systems. The basic characteristics of this process are shown in Table 5.13.

Table 5.13 Characteristics of membrane process (Hillis, 2007).

Process	Mechanism	Exclusion (μm)	Regulated solutes		
			Pathogens	Organics	Inorganics
EDR	C	0.0001	None	None	All
RO	S, D	0.0001	C, B, V	DBPs, SOCs	All
NF	S, D	0.001	C, B, V	DBPs, SOCs	All
UF	S	0.001	C, B, V	None	None
MF	S	0.01	C, B	None	None

Mechanism: *C*, charge; *D*, diffusion; *S*, sieving. Pathogens: *C*, cysts; *B*, bacteria; *V*, viruses; Organics: *DBPs*, disinfection byproduct precursors; *SOCs*, Synthetic Organic Compounds.

Although many factors affect the solute separation by these processes, a general understanding of drinking water application can be achieved by associating the minimum size of solute rejection with membrane process and regulated contaminate. The water quality that is required, determine membrane type. Many regulatory constraints for drinking water can be related to the control of inorganic, organic or pathogenic solutes in the ultimate product. The specific application of membrane processes to drinking water applications is shown in Table 5.14.

Table 5.14 The specific application of membrane processes to drinking water regulation (Hillis, 2007).

US regulation/rule	Membrane process				
	EDR	RO	NF	UF	MF
SWTR/ESWTR	No	Yes	Yes	Yes	Yes
CR	No	Yes	Yes	Yes	Yes
LCR	No	Yes	Yes	No	No
IOC	No	Yes	Yes	No	No
SOC	No	Yes	Yes	Yes	Yes
Radionuclides	yes (-Rn)	yes (-Rn)	yes (-Rn)	No	No
DBPR	No	Yes	Yes	No	No
GWDR	No	Yes	Yes	Yes	Yes
Arsenic	Yes	Yes	Yes	No	No
Sulfates	Yes	Yes	Yes	No	No

CR, Coliform Rule; *DBPR*, Disinfection By-Products Rule; *ESWTR*, Enhanced Surface Water Treatment Rule; *GWDR*, Groundwater Disinfection Rule; *IOC*, Inorganic Rule (Phases I, II, IIA, V); *LCR*, Lead and Copper Rule; *SOC*, Synthetic Organic Chemicals (Base Neutrals and Extractables); *SWTR*, Surface Water Treatment Rule.

Examples of community water quality objectives are total hardness, taste and odor, and color. some community water quality objectives may be classified as secondary standards or those standards which don't affect consumer health. community water quality objectives are shown in Table 5.15.

Ones the treatment objectives are known, potential membrane systems can be determined to achieve these goals. Some general statements about current treatment concerns can be stated:

- Diffusion controlled membrane, e.g., reverse osmosis and nanofiltration, are required to control of inorganic contaminants such as total dissolved solids, total hardness, chlorides, and disinfection byproduct precursors.
- Charge controlled membrane, e.g., electrodialysis reversal can remove total dissolved solids, total hardness, and chlorides.
- Size exclusion-controlled membranes can control particles, turbidity, and cysts.

Case studies are investigations of singular or combined processes with specific goals for production, water quality, and costs. Exact plan documentation is necessary to extend and report pilot production and water quality Documentation of laboratory work should be accompanied by changing custody and quality control. Determination of precision and accuracy of analytical samples in the laboratory or field is a meaningful interpretation of results. A traceable paper trail of sample analysis should be described in project documentation and referenced in a project report. Work and sampling logs should be used for documentation of fieldwork. Work or user logs have daily entries that describe the operation data, mechanical repair or any other activity associated with the operation of the pilot systems. Sample logs are separate from work logs in that specific data is recorded. Typical examples of field data are PH, pressure, flow, and temperature.

Table 5.15 summary of membrane process applications for drinking water rules (Hillis, 2007).

Parameter	**Membrane process**				
	EDR	**RO**	**NF**	**UF**	**MF**
TDS	Yes	Yes	Yes	No	No
TH	Yes	Yes	Yes	No	No
T and O	No	Yes	Yes	No	No
TOC	No	Yes	Yes	No	No
Color	No	Yes	Yes	No	No
FE and Mn	No	Yes	Yes	No	No

Table 5.16 Fouling control by pretreatment system (Hillis, 2007).

Mechanism	Process				
	A/AS	MF/UF	CSF	NH_4Cl	AOC removal
Scaling	+	–	–	–	–
Plugging	–	+	+	–	–
Adsorption	–	+	+	–	–
Bio-fouling	–	–	–	+	+

Productivity is essential to any water treatment facility. Productivity is affected by the design of the membrane process and fouling. Designers can select membranes for specific treatment characteristics. Once selected, a designer can choose the working conditions for that membrane process. The deposition is one of the most important factors affecting productivity. The primary mechanisms of doubling are scaling, plugging, adsorption, and biological growth. The primary means of controlling fouling by mechanism and unit operation are shown in Table 5.16.

General comments can be made regarding pretreatment requirements.

- Scaling control is typically required for all reverse osmosis/nanofiltration membrane systems in either surface or groundwaters and is achieved by acid and antiscalant addition.
- Plugging control is typically required for all reverse osmosis/nanofiltration membrane systems in either surface or groundwaters and is achieved by feed water turbidities and silt density index (SDI) less than 0.2 NTU and 2, respectively.
- Bio-fouling control is typically required for aerobic surface or groundwaters and is achieved by NH_2Cl or addition of other bactericidal agents.
- Organic fouling can occur in the surface water system with total organic carbon (TOC) > 3–6 mg/L and is typically reduced by coagulation, sedimentation, and filtration. However, the significance of organic fouling isn't known (Hillis, 2007).

5.12 Adsorption—reverse osmosis membranes

Reverse osmosis is a process for desalination using a membrane that is water-permeable but essentially impermeable to salt. Reverse osmosis (RO) was used to preconcentrate spent sulfite liquor, and ultrafiltration (UF) to treat bleach plant effluents. Saltwater is in contact with the feed side membrane under pressure. Salt-free and sweet water exits as a low-pressure infiltration. The ability of membranes to separate small solutes from water has been known for a very long time. Reverse osmosis is almost completely limited to water treatment, but an important advance has been the development of similar membrane processes to separate solutes from organic solvents. This technology, perhaps better-called hyperfiltration, is still largely a laboratory-scale operation, but a few industrial plants have been installed.

Although reverse osmosis membranes have been formed and tested using a large number of different materials, cellulosic polymers such as cellulose acetate and cellulose triacetate are the most widely used, however, polyamide and polyuria composite membranes typically have higher water fluxes and salt rejections in reverse osmosis process. Cellulosic polymers are able to withstand higher temperatures and wider pH (pH 4 to pH 11) variation. But polyamide membranes tend to be less chlorine resistant and more susceptible to oxidation as compared to cellulose acetate membranes. The major limitations of the cellulose acetate membrane are that it can only be used in aqueous solutions and that it must be used below about 60°C. Most commonly available RO membranes are divided into two groups as follows:

(a) Asymmetric membrane (skinned)

(b) Thin-film composite membrane

Asymmetric membrane: These membranes have a thin (~100 nm) perm-selective skin layer supported on a more porous sublayer of the same polymer. The dense skin layer determines the fluxes and selectivities of the membrane, whereas the porous sublayer serves only as mechanical support for the skin layer. The thin, dense layer has the ability to block the passage of quite small solute molecules. These membranes are usually prepared by phase inversion methods (Nath, 2017).

Thin-film composite membrane: These membranes include a thin polymeric barrier layer and one or more porous support layers which thin barrier is on top of the support and support layers about ever a different polymer from the surface layer. Here also the surface layer determines the flux, whereas porous backing serves as mechanical support. The barrier layer is extremely thin, thus allowing high water flux. These membranes are generally made by interfacial polymerization, a process in which highly porous membrane, usual polysulfone, is coated with an aqueous solution of a polymer or monomer and then reacts with a cross-linking agent in a nonaqueous solvent (Nath, 2017).

A number of membrane materials and membrane preparation techniques have been used to make reverse osmosis membranes. The target of much of the early work was seawater desalination (approximately 3.5 wt.% salt), which requires membranes with salt rejections of greater than 99.3% to produce an acceptable permeate containing less than 500 ppm salt. Early membranes could only meet this target performance when operated at very high pressures, up to 100 bar. As membrane performance has improved, the operating pressure has dropped to 50–60 bar. Recently, membranes to desalt water feed with salt concentrations of 0.1–0.5 wt.% have been developed. For these applications, membranes are typically operated at pressures in the 10–30 bar range with a target salt rejection of about 99%. With the growth of the electronics industry, the demand for membranes to produce ultrapure water to wash silicon wafers has increased. The feed to an ultrapure water reverse osmosis plant is often municipal drinking water, which may only contain 100–200 ppm dissolved salts, mostly divalent ions. The target membrane performance, in this case, maybe 98%–99% sodium chloride rejection, but more than 99.5% divalent ion rejection. These membranes are operated at low pressures, typically in the 8–12 bar range. Many manufacturers tailor the properties of single membrane material to meet the requirements of different applications. Invariably, a significant trade-off between flux and rejection is involved.

The first high proficiency reverse osmosis membrane discovered was Cellulose acetate. The flux and rejection of cellulose acetate membranes have now been surpassed by interfacial composite membranes. However, cellulose acetate membranes still maintain a small fraction of the market because they are easy to make, mechanically tough, and resistant to degradation by chlorine and other oxidants, a problem with interfacial composite membranes. Cellulose acetate membranes can tolerate continuous exposure of up to 1 ppm chlorine, so chlorination can be used to sterilize the feed water, a major advantage with feed streams having significant bacterial loading (Baker, 2004).

5.13 Economic evaluation/justification

The performance of membrane units can be measured in the field or laboratory by functioning membrane systems in the form of small cells, bench units or pilot plants. Small cells have historically been used to test film characteristics and have not yet been shown to be representative of actual production. A testing protocol for small cells been published. Their results indicated that the maximum contaminant levels of the disinfection byproduct precursors (DBP) in phase II could be met four of the five water tested, the rate of productivity decline was exceptionally high and that dissolved organic carbon (DOC) was diffusion controlled. While the DBP were comparable with previous studies, the productivity results were not. Their results have shown that small cells are suited for membrane screening quality studies but aren't adequate for productivity assessment.

Membrane manufacturers use 4″ × 40″ elements, 2.5″ × 40″ elements or 2.5″ × 20″ elements in pilot studies for plant scale-up. The preliminary results of the information collection rule (ICR) were presented at the water quality and technology conference in Tampa, FL. In excess of 40 membrane studies have been done through the United States to comply with the ICR requirements. The studies usually include flat leaf laboratory work; single element fieldwork, multi-staged pilot manufactories, and full-scale manufactory documentation. The studies have shown that the DBP removal requirements were exceeded for halo acetic acid for either stage 1 or 2 maximum contaminant levels, but that the 95th percentile of trihalomethanes (THMs) exceeded the stage 2 MCL (maximum contaminant level). the higher THMs were attributed to high bromide concentrations and variable bromide rejection (70%—0%) by membranes (Hillis, 2007).

Membrane case studies require that mass transport of water and solute through the membranes be described quantitatively. The equations used to describe flow through a single element are shown in Eqs. (5.1) to (5.7) with reference to the membrane element shown in Fig. 5.29.

Mass transport in pressure-driven membrane processes can be described as convection or diffusion controlled. Models for describing mass transfer in membrane systems have been presented by several investigators. These equations can be used to predict the permeate water quality of any membrane system. However, membrane systems have been shown to produce higher quality than regulative requirements, models, as shown in Eq. (5.8) or (5.9), are necessary for predicting the cost and efficiency of membrane systems from pilot plant data. Eq. (5.8) is used to describe diffusion-controlled mass

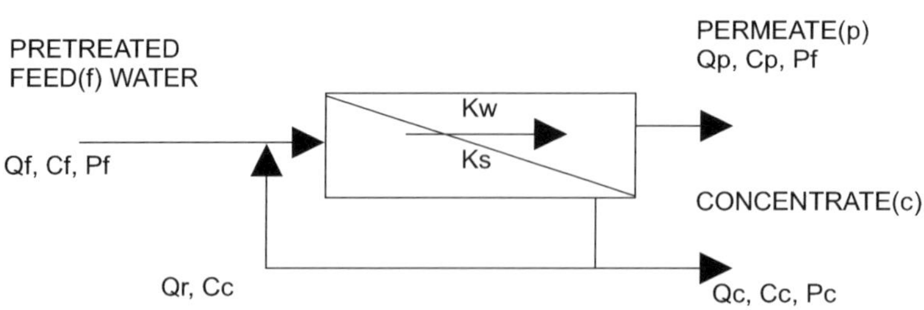

Figure 5.29 Basic diagram of mass transport in a membrane.

transfer, which includes inorganics, i.e., alkalinity, hardness, total dissolved solids (TDS), sodium, chlorides. Eq. (5.9) is used to described sieving controlled mass transfer, which includes total organic carbon (TOC), disinfection by-product precursors (DBP), most synthetic organic chemicals (Base Neutrals and Extractables) (SOCs) and organics in general (Hillis, 2007).

$$F_w = K_w(\Delta P - \Delta\Pi) = \frac{Q_p}{A} \tag{5.8}$$

$$F_s = K_s\Delta C = \frac{Q_p C_p}{A} \tag{5.9}$$

$$R = \frac{Q_p}{Q_f} \tag{5.10}$$

$$Q_f = Q_c + Q_p \tag{5.11}$$

$$Q_f C_f = Q_c C_c + Q_p C_p \tag{5.12}$$

$$\left[\frac{C_s - C_p}{C_f - C_p}\right] = e^{\frac{F_w}{k}} \tag{5.13}$$

$$r = \frac{Q_r}{Q_f} \tag{5.14}$$

$$C_p = \frac{K_s C_f e^{\frac{F_w}{k}}}{F_w\left(\frac{(1-r)(2-2R)}{2+2r-R}\right) + K_s e^{\frac{F_w}{k}}} \tag{5.15}$$

$$C_p = \Phi C_v \tag{5.16}$$

ΔC = Concentration gradient (M/L^3), $\left(\left(C_f + C_c\right)/2 - C_p\right)$,
F_w = Water flux (L^3/L^{2}t),
F_s = Solute flux (M/L^{2}t),
C_f = Feed stream solute concentration (M/L^3),
K_s = Solute mass transfer coefficient (L/t),
Q_f = Feed streamflow (L^3/t),
C_c = Concentrate stream solute concentration (M/L^3),
Q_c = Concentrate stream flow (L^3/t),
C_p = Permeate stream solute concentration (M/L^3),
Q_p = Permeate stream flow (L^3/t),

R = Recovery,
C_s = Solute concentration at the membrane surface (M/L^3),
A = Membrane area (L^2),
r = Recycle ratio,
k = Diffusion coefficient from the surface to the bulk (L^3/L^{2}t),
Φ = Sieving pass coefficient,
ΔP = Pressure gradient (L), $\left(\left(P_f + P_c\right)/2 - P_p\right)$,
$\Delta \Pi$ = Osmotic pressure (L) $\left(\left(\pi_f + \pi_c\right)/2 - \Delta P\right)$,
K_w = Solvent mass transfer coefficient (L^{2}t/M).

5.14 Future opportunities and challenges

As were noted by the US National Research Council, the technological frontiers of the membrane technology should be concerned more in the developing of new membrane materials and the identification of new ways of using perm-selective membranes. New membrane materials to be used is still a big option in the research of this brand-new technology, as most of the researchers are always intending to get a better improvement for this separation process The *Journal of Membrane Science* to serve as a good reference, where lots of the new membrane materials research may be found. For the latter, the membrane-based hybrid system serves as a good example, as it is a combination of conventional unit operations and membrane separation processes, which often results in separation processes that offer considerable advantages over the limited use of either component process. Such advantages may include more complete separation, reduced energy requirement, lower capital cost, and lower production cost (Nath, 2017).

The solution-diffusion model has been less successful at providing a link between the nature of membrane materials and their membrane permeation properties. This link requires an ability to calculate membrane diffusion and sorption coefficients. These calculations require knowledge of the molecular level of interactions of permeant molecules and their motion in the polymer matrix that is not yet available. Only semiempirical correlations such as the dual sorption model or free volume correlations are available. The best hope for future progress toward a priori methods of calculating permeant sorption and diffusion coefficients lies in computer-aided simulations of molecular dynamics, but accurate predictions using this technique are years—perhaps decades—away. Unfortunately, industrially useful membranes have nonuniform tortuous pores and are often anisotropic as well. Current theories cannot predict the permeation properties of these membranes.

5.15 Conclusion

During the last 30 years, the basis of permeation through membranes has become much clearer. This is particularly true for reverse osmosis, gas permeation, and pervaporation, for which the solution-diffusion model is now almost universally accepted and well supported by a body of experimental evidence. The model is popular because it provides simple equations that accurately link the driving forces of concentration and pressure with flux and selectivity. Also, the same set of assumptions can be used to describe permeation in a variety of seemingly different processes. In this chapter, membranes and membrane processes are briefly described. It was also explained how each membrane process utilizes surface adsorption, screening, or carrier separation to separate soluble particles and contamination from feed solution and solvent purification. It can be concluded that membrane adsorption process using both filtration and adsorption processes can simultaneously play an important role in pollutant removal and purification and increase the rate and quality of separation. Because of this, the membrane separation process has many applications and functions in the industry which is increasing day by day.

References

Aptel, P., Challard, N., Cuny, J., & Neel, J. (1976). Application of the pervaporation process to separate azeotropic mixtures. *Journal of Membrane Science, 1*, 271–287.

Bai, H., Zan, X., Zhang, L., & Sun, D. D. (2015). Multi-functional CNT/ZnO/TiO_2 nanocomposite membrane for concurrent filtration and photocatalytic degradation. *Separation and Purification Technology, 156*, 922–930.

Baker, R. W. (2004). *Membrane technology and applications*. John Wiley & Sons.

Bechhold, H. (1907). Kolloidstudien mit der Filtrationsmethode. *Zeitschrift für Physikalische Chemie, 60*, 257–318.

Bernardes, A., Rodrigues, M. A. S., & Ferreira, J. Z. (2016). *Electrodialysis and water reuse*. Springer.

Barão, B. V., Vieira Salla, A. C., Miorando, T., Hemkemeier, M., & Cadore Favaretto, D. P. (2019). Water recovery from dairy rinse water by reverse osmosis: Giving value to water and milk solids. *Resources, Conservation and Recycling, 140*, 313–323.

Cadotte, J. E. (1985). Evolution of composite reverse osmosis membranes. *Materials Science of Synthetic Membranes*, 273–294.

Chen, X., Hu, Y., Xie, Z., & Wang, H. (2018). Materials and design of photocatalytic membranes. In *Current trends and future developments on (Bio-) membranes*. Elsevier.

Choi, H., Elias, S., Dionysios, D., & Dionysiou. (2006). Sol–gel preparation of mesoporous photocatalytic TiO_2 films and TiO_2/Al_2O_3 composite membranes for environmental applications. *Applied Catalysis B: Environmental, 63*, 60–67.
Comerton, A. M., Andrews, R. C., Bagley, D. M., & Yang, P. (2007). Membrane adsorption of endocrine disrupting compounds and pharmaceutically active compounds. *Journal of Membrane Science, 303*, 267–277.
Donnan, F. G. (1995). Theory of membrane equilibria and membrane potentials in the presence of non-dialysing electrolytes. A contribution to physical-chemical physiology. *Journal of Membrane Science, 100*, 45–55.
Doong, S. J. (2012). Membranes, adsorbent materials and solvent-based materials for syngas and hydrogen separation. In *Functional materials for sustainable energy applications*. Elsevier.
Gao, Q., Wang, C.-Z., Liu, S., Hanigan, D., Liu, S.-T., & Zhao, H.-Z. (2019). Ultrafiltration membrane microreactor (MMR) for simultaneous removal of nitrate and phosphate from water. *Chemical Engineering Journal, 355*, 238–246.
Gitis, V., & Rothenberg, G. (2016). *Ceramic membranes: New opportunities and practical applications*. John Wiley & Sons.
Graham, T. (1866). On the absorption and dialytic separation of gases by colloid septa. *The London, Edinburgh, and Dublin Philosophical Magazine and Journal of Science, 32*, 401–420.
He, X., Chen, R., Zhu, X., Liao, Q., Liang, A., Cheng, X., et al. (2016). Optofluidics-based membrane microreactor for wastewater treatment by photocatalytic ozonation. *Industrial & Engineering Chemistry Research, 55*, 8627–8635.
Hillis, P. (2007). *Membrane technology in water and wastewater treatment*. Royal Society of Chemistry.
Horvath, A. G. (1931). Water softening. In *Google patents*.
Ismail, A. F., Khulbe, K. C., & Matsuura, T. (2015). *Gas separation membranes*. Switzerland: Springer.
Jönsson, A.-S., & Gun, T. (1990). Ultrafiltration applications. *Desalination, 77*, 135–179.
Khulbe, K. C., & Matsuura, T. (2018). Removal of heavy metals and pollutants by membrane adsorption techniques. *Applied Water Science, 8*, 19.
Kober, P. A. (1917). Pervaporation, perstillation and percrystallization'. *Journal of the American Chemical Society, 39*, 944–948.
Kuvarega, Alex, T., Bhekie, B., & Mamba. (2016). Photocatalytic membranes for efficient water treatment. In *Semiconductor photocatalysis-materials, mechanisms and applications (IntechOpen)*.
Li, K. (2007). *Ceramic membranes for separation and reaction*. John Wiley & Sons.
Li, B. R., Chen, Z. Y., Wang, J. B., Zhang, Z. L., Fan, Q. J., & Wei, Z. S. (2016). Nitric oxide removal with a Fe-TiO_2/PSF hybrid catalytic membrane bioreactor. *Huan jing ke xue= Huanjing kexue, 37*, 847–853.
Li, X., Jiang, L., & Li, H. (2018). Application of ultrafiltration technology in water treatment. In *IOP conference series: Earth and environmental science* (p. 012009). IOP Publishing.
Li, Y., Sun, J., Liu, L., & Yang, F. (2017). A composite cathode membrane with $CoFe_2O_4$–rGO/PVDF on carbon fiber cloth: Synthesis and performance in a photocatalysis-assisted MFC-MBR system. *Environmental Science: Nano, 4*, 335–345.
Liu, C., Cheng, L., Zhao, Y., & Zhu, L. (2017). Interfacially crosslinked composite porous membranes for ultrafast removal of anionic dyes from water through permeating adsorption. *Journal of Hazardous Materials, 337*, 217–225.
Michaels, A. S. (1971). High flow membrane. In *Google patents*.
Muirhead, A., Beardsley, S., & Aboudiwan, J. (1982). Performance of the 12,000 m^3/d seawater reverse osmosis desalination plant at Jeddah, Saudi Arabia January 1979 through January 1981. *Desalination, 42*, 115–128.
Nath, K. (2017). *Membrane separation processes*. PHI Learning Pvt. Ltd.

Neel, J., Nguyen, Q. T., Clement, R., & Le Blanc, L. (1983). Fractionation of a binary liquid mixture by continuous pervaporation. *Journal of Membrane Science, 15*, 43–62.
Porter, M. C. (1989). *Handbook of industrial membrane technology*.
Reid, C. E., & Breton, E. J. (1959). Water and ion flow across cellulosic membranes. *Journal of Applied Polymer Science, 1*, 133–143.
Scott, K., & Hughes, R. (2012). *Industrial membrane separation technology*. Springer Science & Business Media.
Strathmann, H. (2004). *Ion-exchange membrane separation processes*. Elsevier.
Strazik, W., & Perry, E. (1973). Process for the separation of styrene from ethylbenzene. In *Google patents*.
Uhlhorn, R. J. R., Keizer, K., & Burggraaf, A. J. (1992). Gas transport and separation with ceramic membranes. Part II. Synthesis and separation properties of microporous membranes. *Journal of Membrane Science, 66*, 271–287.
Wang, G., Chen, S., Yu, H., & Xie, Q. (2015). Integration of membrane filtration and photoelectrocatalysis using a TiO_2/carbon/Al_2O_3 membrane for enhanced water treatment. *Journal of Hazardous Materials, 299*, 27–34.
Zhang, X., Wang, D. K., Schmeda Lopez, D. R., & da Costa, J. C. D. (2014). Fabrication of nanostructured TiO_2 hollow fiber photocatalytic membrane and application for wastewater treatment. *Chemical Engineering Journal, 236*, 314–322.
Zhang, H., Xie, Q., Chen, S., Zhao, H., & Zhao, Y. (2006). The removal of sodium dodecylbenzene sulfonate surfactant from water using silica/titania nanorods/nanotubes composite membrane with photocatalytic capability. *Applied Surface Science, 252*, 8598–8604.
Zhao, H., Chen, S., Xie, Q., Yu, H., & Zhao, H. (2016). Integration of microfiltration and visible-light-driven photocatalysis on g-C_3N_4 nanosheet/reduced graphene oxide membrane for enhanced water treatment. *Applied Catalysis B: Environmental, 194*, 134–140.

CHAPTER SIX

Sorption based on the geometry design and equipment structure

6.1 Introduction

The design and development of a dynamic system for the sorption processes requires considering different factors such as adsorbent/adsorption type and its distribution, gas/fluid flow rates, solute concentration, and the geometry of the contactor. As can be seen in the adsorption process, the breakthrough curve which determines the dynamic adsorption process is a function of both the operating conditions and bed geometry (Moreno-Pérez et al., 2016).

Fixed bed columns can be designed for different processes owing to their flexibility and low energy requirements. For some cases, e.g., the wastewater treatment process, to the reduction in the costs associated with the design, construction, more evaluations should be considered. The adsorption of the pollutants from a wastewater flow with high flow rate requires adsorption columns with large bed heights to achieve the desired standards. In this regard, dynamic systems of multiple columns can be used with an arrangement of two or more fixed beds. However, the cost and size of the separation system are critical parameters that restrict the number of beds operated at such configuration. Accordingly, compact devices as a result of the modifications in the geometry are proposed to reduce the dimension of sorption systems, while they have less installation, maintenance, and operation costs.

6.2 Sorption processes using fixed beds

Fixed beds are well-knows devices which are used for many separation processes such as absorption, stripping and distillation. In the fixed-bed columns, the phase flow distribution through the bed has a great role in process performance. Furthermore, the choice of the packing is an important factor that should be considered during the design process. An appropriate packing

Intensification of Sorption Processes
ISBN: 978-0-12-821411-4
https://doi.org/10.1016/B978-0-12-821411-4.00003-1

provides a large surface, while it has low resistance to phase flows. Creation of a uniform liquid and gas distribution across the bed is another parameter which should be considered for choice a packing. In general, the packings are classified into two types: packing with regular geometry (structured packings) and random packing.

6.2.1 Adsorption process using fixed beds

There are different techniques such as batch, continuous moving bed, continuous fixed bed, continuous fluidize bed, and pulsed bed, which are used for adsorption process through the contact between adsorbate and adsorbent. Each technique has some advantages and disadvantages which are presented in Table 6.1. As can be seen, the fixed-bed column is more preferable and industrially feasible for adsorption of different pollution. The breakthrough curves are used for evaluation of the adsorption process performances, while the results could be provided via the pollutant concentration versus time profile. The axial dispersion, film diffusion resistance, intraparticle diffusion resistance and sorption equilibrium with the sorbent are accepted phenomena which justifying the adsorption mechanism.

The relation between the nature of breakthrough curves and fixed-bed adsorption is distinguished through the mass transfer zone (MTZ) or primary sorption zone (PSZ). This relation is presented in Fig. 6.1. The feed (contaminant solution) is inserted from the top of the column, while the adsorbate is adsorbed by the upper few layers of the fresh adsorbent during the initial stage of the operations. This event is owing to the higher value of adsorbent and small levels of adsorbate available at these upper layers. In this condition, the adsorbate is readily escaped in the lower strata of the bed and no adsorbate run off from the adsorbent at the first stage. Furthermore, primary adsorption zone or MTZ is created near the top or influent end of the packed column, whereas the concentration of adsorbate (C) is zero, and accordingly the ratio of effluent and initial concentration (C/C_0) is zero. Afterward, the upper layer of adsorbent is gradually saturated, with feeding the adsorbate into the column, which becomes adsorbent less efficient progressively. Therefore, the PSZ also travels descending to fresher or unadsorbed part of adsorbent in the column.

Moreover, with movement of this zone, tendency is that more and more adsorbate comes out in the effluent as per points C_1/C_0, C_2/C_0, C_3/C_0 and C_4/C_0. The movement of this zone is mainly increasing with

Table 6.1 The advantages and disadvantages of various sorption processes (Patel, 2019).

Particular	Batch sorption	Continuous fixed-bed sorption	Continuous moving bed sorption	Continuous fluidized bed sorption	Pulsed bed sorption
Introduction	Adsorbent and adsorbate are well mixed in diluted solution at constant volume in well-mixed system	Fixed-bed system consists of an adsorbent in which adsorbate is continuously flowed through a bed of adsorbent at constant rate	Continuous moving bed sorption is steady-state system, where both adsorbent and adsorbate are in motion, and bed of adsorbent section remains at constant, but not in equal condition	In this sorption, adsorbate is in contact with fluidized bed of adsorbent with sufficient or insufficient flow	In pulsed bed sorption, adsorbate is contacted with same adsorbent in bed, until desired results are not achieved
Features	Very easy and cheap technique	Very easy and cheap technique	Complicated and very expensive technique	Complicated and very expensive technique	Very easy and cheap technique
	Most of the researchers are using this technique to analyze feasibility of adsorbent-adsorbate system	Used for higher quantity of wastewater having higher pollution load	As adsorbent is continuously replaced and fresh adsorbent is constant contact with adsorbate	Used for higher quantity of wastewater having higher pollution load	It is very easily controlled automatically operated system
		Also, widely used for industrial purpose, because the adsorbate is continuously in contact with a given quantity of fresh adsorbent in fixed-bed column system		Also, applicable for industries because it allows rapid mixing of adsorbent—adsorbate and also, adsorbate is continuously flow automatically with controlled operation and easy handing	Also, it required lower dosage of adsorbent. This type has an advantage of better utilization of adsorbent because the adsorbents were kept for regeneration as soon as the adsorbent gets saturated

(*Continued*)

Table 6.1 The advantages and disadvantages of various sorption processes (Patel, 2019).—cont'd

Particular	Batch sorption	Continuous fixed-bed sorption	Continuous moving bed sorption	Continuous fluidized bed sorption	Pulsed bed sorption
Disadvantages	Used for small quantity of wastewater having minimum pollution load; therefore, this operation is that it is scarcely found in the majority of practical (industrial) applications	The problems associated with this sorption are adsorbent attrition, feed channeling, and nonuniform flow of adsorbent particles	The large amount of adsorbent is required to complete sorption	Flow of adsorbate is not measured with large deviation from plug flow and bubbling or feed channeling, which leads to insufficient contact of adsorbent-adsorbate	Used for small quantity of wastewater having minimum pollution load especially lower suspected solid
	Adsorbent is removed from the system by simple filtration method	Forceful interaction is conducted in continuous fixed-bed systems to reduce space and time. As a result, it is difficult to carry out a priori design and optimization of fixed-bed columns without a quantitative approach	Continuous regeneration of adsorbent and adsorbent storage is essential	The rapid mixing of adsorbent-adsorbate system leads to nonuniform residence time	Adsorbent is not unfilled in normal operations

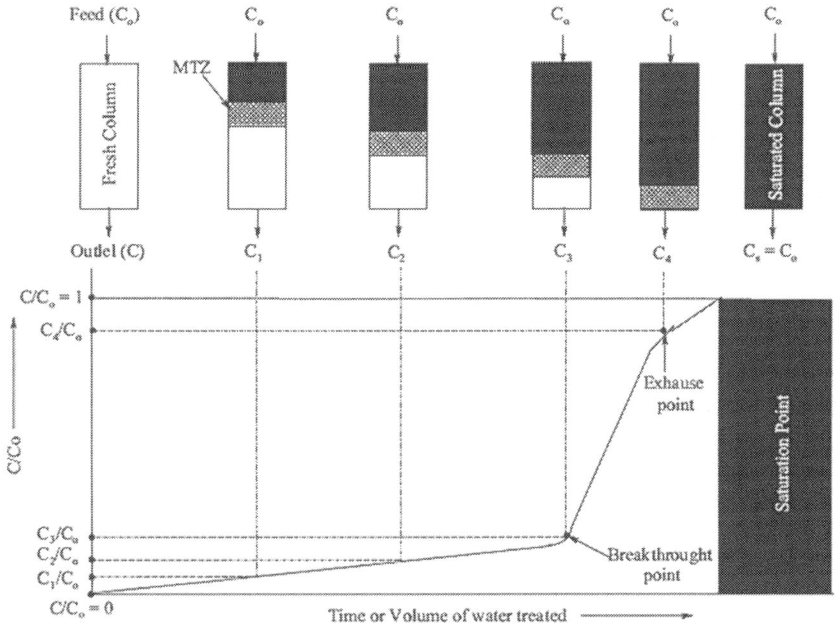

Figure 6.1 The breakthrough curve and MTZ (Patel, 2019).

increasing initial concentration compared to linear velocity of the feed. During the operation, at a specific time (Cs), the column is completely saturated and consequently, the adsorption is stopped, while in this state the ratio of C/C_0 is 1. For most adsorption processes using a fixed column, the breakthrough curves give a characteristic "S" shape, however, the degree of steepness is different for each process.

Table 6.2 provides applications of fixed-bed column technology toward the removal of different contaminants.

6.2.2 Absorption process using fixed beds

Another application of fixed beds is their use in the absorption processes. Chemical absorption which is named reactive absorption, is the process where gas is absorbed by a liquid phase with a combination of reaction and absorptive mass transport. Table 6.3 presents some applications of fixed-bed columns for absorption processes.

Table 6.2 Application of fixed-bed columns for adsorption of different species.

Phase adsorption					
Liquid phase			Gas phase		
List	Samples	Refs.	List	Samples	Refs.
Metal ions	Pb^{2+}, Cu^{2+}, CrO_4^{2-}	Jiang, Qing-Da Xiao, Zhai, and Shi (2019)	Organic solvent vapors	Benzene	Hussein and Ahmed (2016)
	Ni^{2+}, Pb^{2+}, Cu^{2+}	Mahdi, Yu, and El Hanandeh (2019)		n-butyl acetate	Sui et al. (2019)
	Pb^{2+}	Xiong, Zhang, Jun-Xia, and Chi (2019)		p-xylene	Sui et al. (2019)
	Cu^{2+}, Co^{2+}, Ni^{2+}	Xavier et al. (2018)		o-xylene	Fakhfakh, Dammak, and Benzina (2018)
Dissolved organic species from water supplies	Dissolved organic nitrogen	Mohtadi, James, and Davis (2020)	Removing toxic agents	Formaldehyde	Zhang et al. (2020)
	Dissolved organic nitrogen	Valencia et al. (2020)			
	Dissolved organic carbon	Sounthararajah, Loganathan, Kandasamy, and Vigneswaran (2016)			
Odors and taste	Geosmin and MIB	Huang et al. (2019)	Gases separation	H_2S capture	Iliuta and Larachi (2018), de Oliveira et al. (2019)
	Geosmin and MIB	Yuan, Huang, Nie, and Hofmann (2020)		CO_2 capture	Jaberi, Mosleh, and Dashtian (2020); Jaberi, Mosleh, Dashtian, and Salchi (2020), Babar et al. (2019)
Decolorization	Remazol red	Kurade, Waghmode, Xiong, Govindwar, and Jeon (2019)	Toxic volatile organic compounds	Pentane, hexane, heptane, benzene, butanone	Wu et al. (2016)
	Methylene blue	Li, Fan, and Sun (2020)			
	Reactive blue—13	Awasthi and Datta (2019)			

Table 6.3 Applications of fixed bed columns for absorption processes.

System	Packing type	Absorbent	Refs.
Removal of CO_2 from a gas mixture (N_2 and CO_2)	Raschig rings	Imidazolium based ionic liquids and MEA solution	Ziobrowski, Krupiczka, and Rotkegel (2016)
Removal of CO_2 from a gas mixture (air and CO_2)	Raschig rings	Hydrochar supported ZIF-8 MOF nanofluid	Jaberi, Mosleh, and Dashtian (2020), Jaberi, Mosleh, Dashtian, and Salehi (2020)
SO_2 removal from simulated flue gas	Berl saddle	AMP and disodium hydrogen phosphate	Rahmani, Mowla, Karimi, Ali and Rahmatmand (2015)
Removal of CO_2 from a gas mixture (air and CO_2)	Ceramic raschig rings and berl saddles	AMP and MEA	Kang, Kim, Kim, Kang, and Oh (2015).
Removal of CO_2 from a gas mixture (air and CO_2)	Raschig rings	CC-hydrochar/ ZIF-67 nanofluid	Jaberi, Mosleh, and Dashtian (2020)

6.3 Sorption processes using helical coil-packed-bed columns

This section evaluates the feasibility to improve the dynamic sorption systems performance using helical coil configurations (Fig. 6.2). Helical coils have applied successfully in various industries such as power generation, nuclear industry, refrigeration, and food industry. According to the analogy between heat and mass transfers, the helical coil geometry for sorption columns may provide some operational benefits compared to straight bed columns. The helical coil bed configuration presents a compact structure which takes up less space. This property is important in the view of economic and safety. Furthermore, the flow in curve pipes is more stable than flow in straight pipes.

The application of the helical coil configuration can improve mass transfer rates in different processes such as reverse osmosis. However, few studies have been done for this configuration related to the sorption systems. The

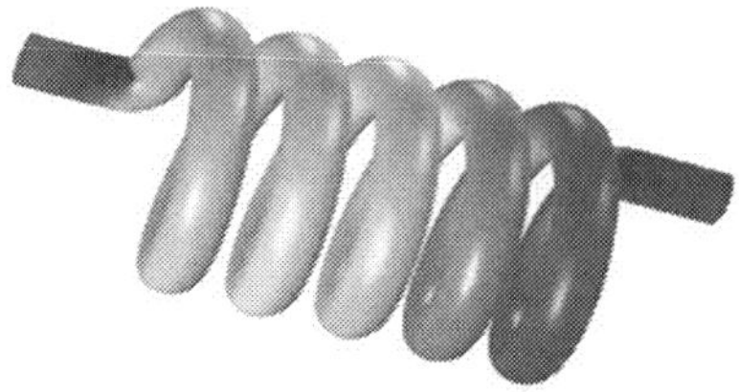

Figure 6.2 Helical coil configuration.

performance evaluation of a helical coil-packed-bed columns for Zn^{2+} adsorption on bone char has been carried out (Moreno-Pérez et al., 2016). The breakthrough curves have been achieved using helical coil columns with different characteristics and a comparison has been done with respect to the results of straight fixed-bed columns. Findings indicated that the helical coil adsorption columns provide an equivalent removal performance than that obtained for the traditional packed-bed columns. It should be noted that using a compact configuration corresponds to the more economic and safety process, hence the helical coil adsorption columns could be considered as a good alternative. The coil diameter, number of turns, and feed flow are key parameters which should be optimized to achieve the best performance in packed-bed geometry. In general, it seems that helical coil columns provide a feasible configuration for large-scale sorption processes with high flow rates where a significant reduction in the process size can be achieved without compromising the process performance.

6.4 Static mixers

The static mixer is a mixer device which has not any moving part (Chemtech, 1997). This device is also called the stationary mixer or in-line mixer. The static mixer is applied to improve the distributive mixing uniformity in the radial direction, while it is placed inside the adaptor between a screw and a die, subdivides the flow into multiple layers, reorients, and mixes the layers (Fig. 6.3).

The fluid flow near the center is directed to the wall and vice versa, mixing and reorienting the layers. Since the static mixer cannot pump the fluid owing to have not any moving part, the fluid should be moved through the static mixer by the screw, which leads to occurring a pressure drop during the static mixer path. They are applied to mix liquid and gas streams, disperse gas into liquid or blend immiscible liquids, while the required energy for mixing of fluids supplies from the pressure drop across the mixer's elements.

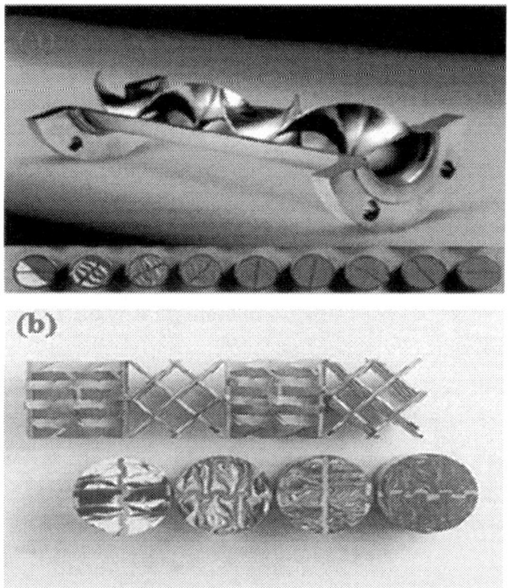

Figure 6.3 Two different type of static mixers: (A) Kenics KM mixer, (B) Koch SMX mixer.

Up to now, different types of static mixers have been designed and used for different applications. One of them is the Kenics KM mixer (Fig. 6.3A), which has a simple structure, including of a series of right-handed helical elements connected to left-handed helical elements placed in a tubular adaptor. In this design, each element is rotated 90 degrees from the preceding element, while the diameter of the tubular adaptor is much smaller than the diameter of the extruder, and each element divides the stream into two layers and rotates the layers 180 degrees. Another design is the Koch SMX mixer (Fig. 6.3B) which consists of a series of mixing elements with an open lattice geometry of intermeshing guide bars or blades located in a large tubular adaptor (Chung, 2019). Each element is rotated 90 degrees from the preceding element, whereas the diameter of the tubular adaptor can be larger than the diameter of the extruder. The special geometry of the bars or the vanes creates flow channels, which intersect each other and make a stream from the center to the wall of the adaptor and vice versa. Although the static mixers can improve the temperature uniformity and the distributive mixing in the radial direction, they cannot mix the stream in the axial direction. Furthermore, they cannot enhance the pressure stability, while the shear rates through them are often low. Despite unique advantages such as small volume, low maintenance, simple installation, and cleaning and excellent reliability, static mixers suffer from some limitations including possible dead-spots, higher head pressure, and longer residence time. An efficient static mixer not only has a low-pressure drop and small volume but also creates many layers, reorient, and mix them (Chung, 2019).

Generally, two main types of static mixers are used at industrial-scale: housed-elements type, and plate-type mixer. Housed-elements type is made of several elements located in a housing. When the fluid flow moves through the mixer, the static elements continuously blend the fluid materials. The fluids properties, the inner diameter of the tube, and the number of elements are some of the design parameters that affect the degree of mixing. Helical elements can simultaneously create patterns of flow division and radial mixing which leads to reduction or removal of the radial gradients in temperature, velocity, and material composition (Fig. 6.4).

In the plate-type static mixers, mixing is performed through the vigorous turbulence in the fluid flow. The corrugated plate geometry is applied to mixing low viscosity liquids, blending gases, dispersing immiscible liquids, and creating gas—liquid dispersions. This operation is performed with a very high degree of mixing in a short length. The number of layers, corrugation angle, and spacers between mixing elements are important design factors that can be used to obtain an excellent performance, which cannot achieve via traditional technologies.

Static mixers have various applications. They have been used successfully in many industrial activities including wastewater treatment systems, acid—base neutralization process, oxidation and bleaching, gaseous reactant

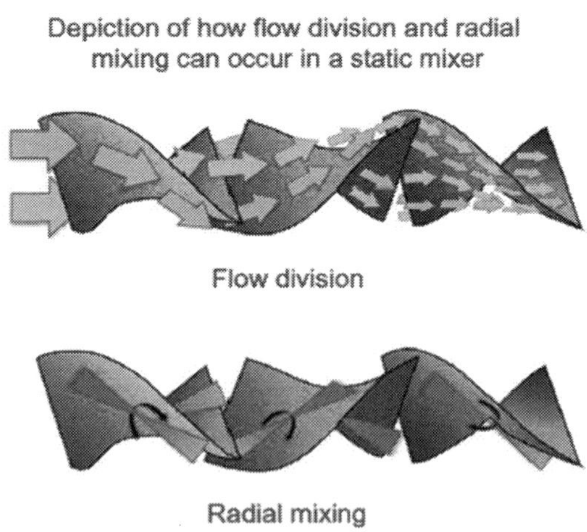

Figure 6.4 Flow patterns is the static mixer (Dimian, Bildea, & Kiss, 2014).

blending, blending of multicomponent drugs, production of the fertilizer and pesticide, steam injection systems, organic-aqueous dispersion operation, oil and gas related-processes, and chemical processing (Dimian et al., 2014).

Kenics static mixers can be used as an efficient photocatalytic reactor for intensification of degradation processes. They are coated with photocatalysts or operate at the suspension state. The performance of the static mixers has been evaluated for photocatalytic degradation of the antibiotic oxytetracycline in aqueous solution, while it was constructed in a borosilicate tube coupled to a compound parabolic collector (Díez et al., 2018). The geometry of the Kenics static mixer creates vigorous mixing even under laminar flow condition. Furthermore, using a compound parabolic collector to receiving sunlight ensures that the entire photocatalyst surface area receives front-side illumination. Kenics static mixers can be used as catalyst support for heterogeneous photocatalysis in tubular photoreactors with compound parabolic collectors (Fig. 6.5).

In general, this configuration intensifies photocatalytic degradation efficiency via two mechanisms:

- Improving the mass transfer by providing intense mixing even under laminar flow regime which leads to enhancing the diffusion of pollutants/reactants to reach the liquid-photocatalyst interface and the active surface sites inside the catalyst film.
- Improving the photon transfer, since the entire photocatalyst surface was receiving front-side illumination.

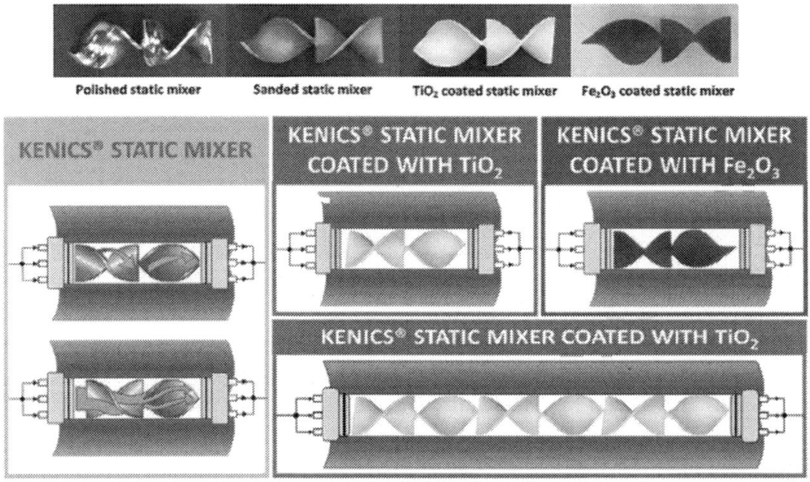

Figure 6.5 Photocatalytic static mixer reactor (Díez et al., 2018).

NETmix is another type of the static mixers which has been used as a reactor. NETmix reactor has shown good performance for homogeneous photo-fenton treatment of natural waters contaminated with organic pollutants (Lima et al., 2016). Findings for degradation of three antibiotics presenting high bacterial resistance including ciprofloxacin, sulfamethoxazole and trimethoprim reveals that complete conversion was achieved, which means that NETmix reactor configuration is a great alternative for conventional technologies.

Although NETmix coupled with compound parabolic collectors can operate in similar photocatalytic degradation reactor volumes, at high flow rate operations, a comparison of performances is required to the evaluation of each equipment (Lima et al., 2017). Despite acceptable achieved degradation efficiency, the scale-up of such types of reactors are difficult owing to their tubular structure, while they require a reasonable capital cost and generate a moderate amount of heat during operation.

6.4.1 Static mixers for gas-liquid separation processes

Static mixers can act as an efficient contactor instead of the bubble column and mechanically stirred vessels. For instance, the enhanced dispersion using a helical static mixer in a vertical pipe at turbulent gas—liquid flow conditions has been investigated, while the dispersive mixing was quantified upstream and downstream of three, six and nine helical static mixer elements in terms of bubble size distribution and gas holdup (Rabha, Schubert, Grugel, Banowski, & Hampel, 2015). The schematic of the used static mixer is shown in Fig. 6.6. P1, P2 and P3 are the pressure drop measurements points, whereas all the dimensions are in mm.

The analysis of the obtained results lead to the following achievements (Rabha et al., 2015):

- The number of static mixer elements at lower slip (fluid velocity) ratio is a key parameter that impacts on the bubble breakup.
- Gas holdup and specific interfacial area depend on the number of static mixer elements (mixer length) as well as on the slip ratio. The effect of increasing the number of static mixer elements, which increases gas holdup and specific interfacial area, is most significant at higher liquid velocities.
- The number of static mixer elements and the liquid velocity has a greater effect on the pressure drop than the gas velocity.

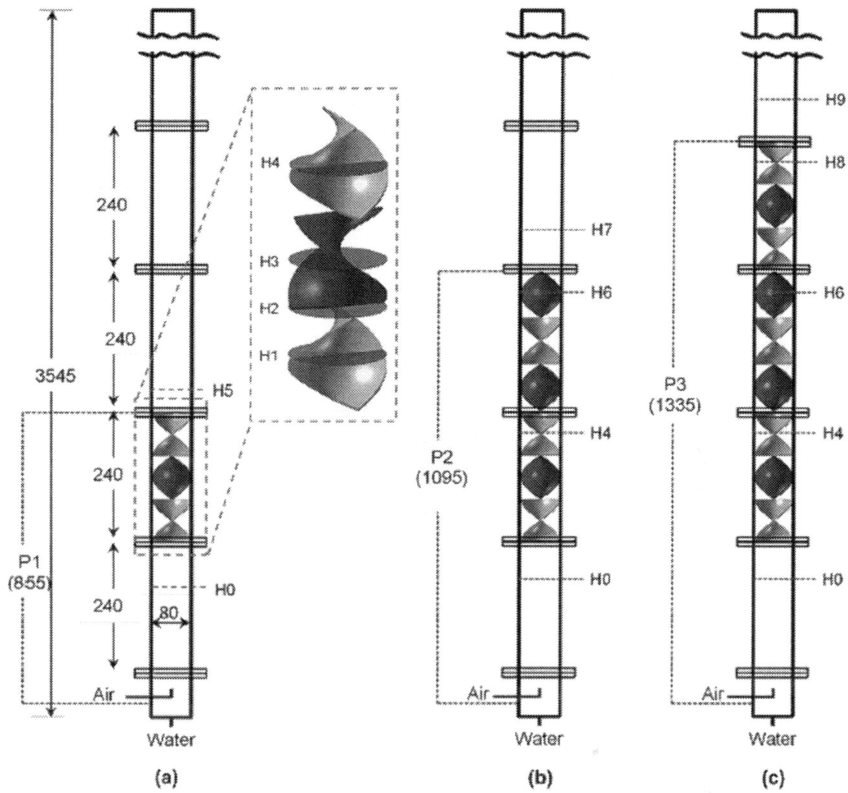

Figure 6.6 Schematic of the static mixer for gas–liquid separation processes. (A) three, (B) six, and (C) nine elements (Rabha et al., 2015).

- The power dissipation per unit mass of the liquid increases with the number of the static mixer elements. However, the effect of liquid velocity on the power dissipation is more pronounced than that of the number of static mixer element and gas superficial velocity.

6.5 Oscillatory baffled devises

The oscillatory baffled column (OBC) is a novel device that can be used for the intensification of the mixing process. This technology improves the performance of the different gas–liquid systems. The high performance of the OBC is related to the eddy mixing in the periodic baffles or constrictions delivers which provides a good degree of radial mixing and secondary flow. In this condition, a very effective controlling of the bubble/drop size distribution is achieved in the column, which enhances the contact between immiscible phases (Pereira, Sousa, Alves, Mackley, & Reis, 2014). Typically, Table 6.4 provides the application of the OBCs for intensification of the O_2 and CO_2 dissolution in water. As can be seen, high mass transfer coefficients have been obtained achieved for different gas—liquid system. One of the beneficial equipment that operated based on the OBC concept, is oscillatory baffled reactor (OBR). OBRs (Fig. 6.7) are an efficient alternative for replacing with conventional plug-flow reactors. This device is designed in such a way that the flow encounters resistances (baffles) located at equally spaced intervals along the flow path, and an oscillatory motion is imposed over the net flow by means of a secondary pump/actuator. This type of reactor provides an excellent opportunity for reactions that require long residence times, especially where a conventional plug-flow reactor would have an impractical length (Baldea & Edgar, 2018).

Oscillatory flow reactor (OFR) is another device that takes unique advantages of the oscillatory flow. This device has an excellent performance to intensify multiphase mixtures with high efficiency. The fluid is oscillated along the reactor equipped with periodic constrictions (baffles, with variable geometric configuration), which forces the movement of the liquid from the walls to the center of the reactor. In this condition, small bubbles (bubble size reduction up to ~ 1 mm with oscillation) are created which provide a high contact time between the phases and also intensifying mixing process, consequently enhance the mass transfer rate. OFRs are applied successfully for intensification of the mass transfer between oxygen and ozone in water which findings indicated that mass transfer rates are several orders of magnitude higher than in conventional contactors. Other types of OFRs can be designed for different application. For instance, using smooth periodic constrictions (SPC) with specific dimensions, novel OFRs can be constructed which have higher performance than that obtained with other ordinary OFR configurations. This type of OFRs is used for intensification of the oxygen-water mass transfer, in the presence and absence of solids and crystallization of proteins and active pharmaceutical ingredients. This technology is an alternative to intensify the ozonation process in water and wastewater treatment systems. The obtained results for intensification of the ozone-water mass transfer using continuous mode using a macroscale oscillatory flow reactor provided with smooth periodic constrictions have been interesting. The effects of oscillation amplitude, frequency, and gas and liquid flow rates on the mass transfer coefficient (k_La) have been evaluated which results revealed that k_La increases with oscillation amplitude,

Table 6.4 Application of the OBCs for intensification of the O_2 and CO_2 dissolution in water (Pereira et al., 2014).

OBC	gas–liquid system	i.d. [mm]	Q_{gas} [vvm]	U_G [mm s^{-1}]	d_o [mm]	α [%]	K_La [h^{-1}]
Batch single-orifice OBC	Air–fermentation media	50	0.5	3.2	20[a]	16	~90–450
Batch reciprocating plate baffled column	Air–water (self-aerating)	190	n/a	n/a	10–50	7–31	~0–23
Batch reciprocating plate baffled column	Air–water	16.6	n/d	5–15	7.8	46.6	180–2880
Batch single-orifice OBC	Air–water	50	0.05–0.2	1.1–4.3	24	23	~0–144
Continuous single-orifice OBC	Pure CO_2–water	94	1.3–3.6	26–72	50	28	~8–100
Continuous, single-orifice meso-OBC	Air–water	4.4	0.064	0.37	1.6	14	~0–576

[a]Authors reported a baffles width/diameter of 30 mm, so it was assumed an open orifice diameter of 20 mm in the calculations; (n/d) not disclosed by the authors; (n/a) not applicable/available.

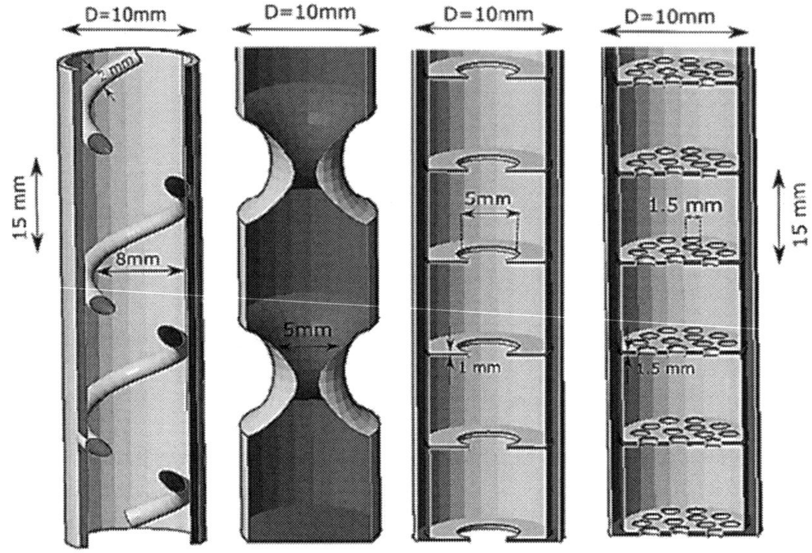

Figure 6.7 Schematic diagram of different OBR configurations (Ahmed, Phan, & Harvey, 2018).

frequency, and the gas flow rate. Furthermore, findings have been indicated that no significant changes in k_La were observed over a wide range of liquid flow rates. In general, using different OFR-SPC configurations tested, the higher k_La values were observed than that achieved using conventional devices. The obtained k_La values up to 35 times higher than those obtained in a conventional bubble column. Furthermore, the enhancement on the mass transfer rate through the oscillatory movement was more pronounced as the gas (ozone) flow rate decreases and liquid (water) flow rate increases, which is another advantage of this system when looking for large-scale implementation of the ozonation process in water treatment. Degradation of the fluoroquinolones in different wastewaters using ozonation process have been revealed that OFR-SPC is an interesting reactor alternative for ozone-driven water treatment (Graça, Rita Bravo Lima, Adrián, & Ferreira, 2020).

The performance of a compact multiorifice oscillatory baffled column (MOBC) as an ozonation contactor has been shown a great efficiency at a very short liquid contact time under continuous operation (Lucas, Reis, & Puma, 2016). Moreover, this system doesn't require further gas—liquid contacting equipment downstream. The performance evaluation of the MOBC ozonation reactor against the conventional equipment has been revealed a

20% increase on the rate of p-hydroxybenzoic acid (p-HBA) degradation, 75% increase in the rate of mineralization of p-HBA per mole of ozone consumed, and 4.5−5.0-fold increase in the rate of mineralization of p-HBA per mole of ozone supplied (Lucas et al., 2016).

6.6 Microfluidic devices

The microfluidics devices (Fig. 6.8) have very small dimensions (e.g., micron) which are used for some purposes such as reduce characteristic diffusion times and support intensified surface processes (e.g., reactions) (Tonkovich & Daymo, 2017). These unique devices have great potential to intensify different chemical processes, making them safer, cleaner and more efficient, which is mainly due to the increased surface to volume ratio.

Microreactors are intensified reactors which are designed based on the characteristics of the microfluidic devices. The better controllability of highly exothermic reactions and narrow residence time distributions are unique merits which are obtained using these reactors owing to the enhanced heat transfer and mass transport. The low reagent consumption is another advantage of microreactors which not only enables the process to be cost-efficient but also causes that limited risk will exist during the usage of hazardous substances.

Microchannel (MC) reactor is an alternative technique for intensification of the absorption process, which provides more than two orders of intensification factor in compared with conventional equipment (Tay, Lau, Lai, Shariff, & Wang, 2019). Using MC_S, the mass transfer coefficient can be enhanced up to 100 s^{-1}, which is about 100 times higher than traditional equipment. MCs have some unique merits such as a minimum additional energy requirement, no moving part, and high chemical stability. Moreover,

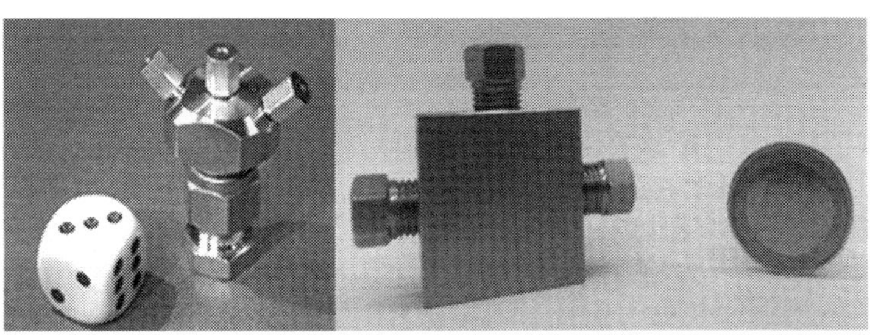

Figure 6.8 HPIMM micromixer from Fraunhofer ICT-IMM (on the left) and tee-mixer (on the right) (Domínguez & Gamse, 2016).

MCs provide high heat transfer coefficient value, which is a crucial aspect for the bulk CO_2 capture process. The simulations of CO_2 absorption using MC_S indicate that the mass transfer coefficient is 5 times higher than the conventional packed-bed column (Firuzi & Sadeghi, 2018). Furthermore, the CO_2 absorption under high-pressure condition using MCs show that the mass transfer coefficient is up to 100 times higher than the traditional columns (Yao et al., 2017). These findings are a great step for design novel equipment toward natural gas purification process. The evaluations for a falling film microreactor system indicate that the interfacial area can be up to 40,000 m^2/m^3 using MCs, which is much higher than the traditional devices by two orders of magnitude (Jähnisch et al., 2000). Generally, the high performance of CO_2 capture process provided by microreactors is owing to the huge interfacial area for mass and heat transfer, improved process control and low material holdup (Yao et al., 2017). The evaluation of the process intensification that can be achieved in the microchannel reactor (Fig. 6.9) has been done by comparing the mass transfer coefficients obtained with conventional equipment (Yao et al., 2017). Findings indicate that the microreactor has much higher performance than conventional equipment. For physical absorption, mass transfer coefficient with a range of 0.45−4.32 s^{-1} is generally larger by one to two orders of magnitude, while for chemical absorption, the reactor performance is higher by one to three orders of magnitude with mass transfer coefficient ranging from 1.59 to 90 s^{-1}.

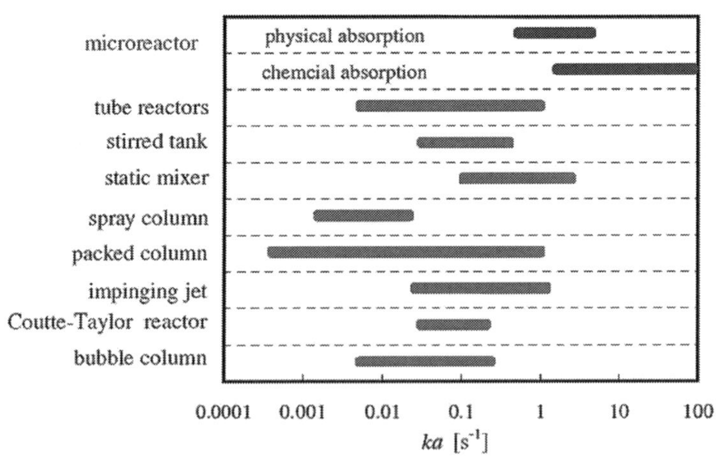

Figure 6.9 Evaluation of the mass transfer coefficients using different equipment (Yao et al., 2017).

The C_2H_4 separation from C_2H_4/C_2H_6 mixture using aqueous $AgNO_3$ solutions in microchannels has been done, which results indicate that the obtained mass transfer coefficients were 3.1–85.1 s^{-1} (Mi, Yao, & Chen, 2018). These values are two orders of magnitude higher than in conventional large-scale contactors. In fact, the gas–liquid absorption is intensified by microreactors owing to the huge interfacial area and short radial diffusion distance which leads to enhancement of the overall mass transfer coefficients.

Distillation and extraction are other processes which can be carried out using microfluidic devices. For example, the supercritical extraction of a substance which includes the continuous separation of a supercritical fluid and a liquid at high pressure can be done via microdevices. In this regard, the extraction of vanillin from an aqueous stream, using $scCO_2$ as solvent has been evaluated at pressures from 80 to 110 bar, which obtained results have been more desirable (Assmann, Kaiser, & von Rohr, 2012). Findings show that applying microextractors in series provides higher purities of the extract with a multistep process (Assmann et al., 2012). The fast and safe measurement of distribution coefficients, as well as the testing of the flammable supercritical solvents, are some goals that can be achieved using microdevices.

Microreactors can be used for the separation of liquid drops or slugs from a second immiscible liquid (Scheiff, Mendorf, Agar, Reis, & Mackley, 2011). In general, the organic/aqueous phase separation can be carried out on the microscale by exploiting wettability differences between the liquids on hydrophilic or hydrophobic surfaces. Rapid phase separation is an important aspect for enhancing yields which can be achieved through the microreactors. The resolution of fine dispersions or emulsions with droplet sizes down to 10 mm can be carried out through the microstructured separation nozzles, in which the driving force for phase disengagement is the lift force induced by the shear stress on the surface of the dispersed phase. The small parallel microchannel (in the range of 100–200 mm) is an appropriate choice, which can be applied instead of membranes to exploit capillary forces for phase separation purposes (Scheiff et al., 2011). Furthermore, the wall induced dielectrophoretic lateral migrations of oil droplets and Janus droplets under different strengths of electrical fields has been done using microchannels, which result show that the separations of oil droplets by size, separations of Janus droplets by size, and separation of mixtures of Janus droplets and oil droplets with the same size can be obtained successfully via adjusting the voltages applied to the microchannel (Li & Li, 2017).

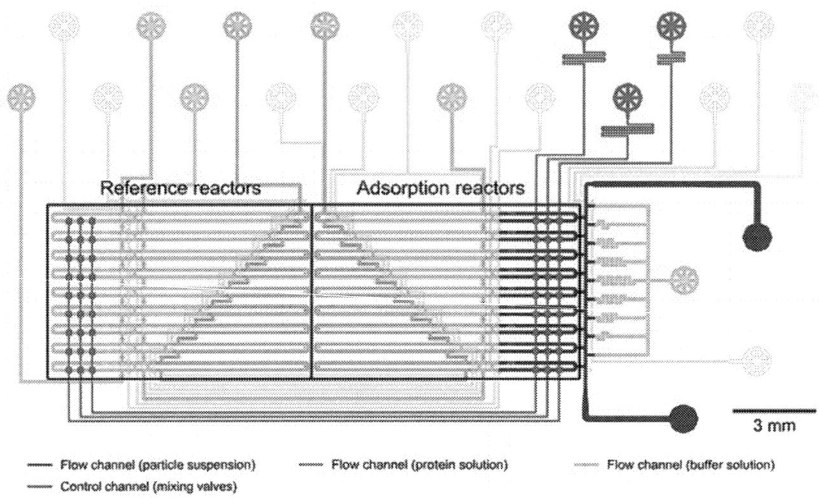

Figure 6.10 The schematic of a microfluidic reactor for protein adsorption (Rho, Hanke, Ottens, & Gardeniers, 2017).

Microfluidic devices can be used for protein adsorption owing to the large value of surface area relative to volume (Fig. 6.10). The protein adsorption process is the first phase in some biological processes such as the attachment of cells to bioengineered surfaces, the coagulation of blood, and the response of the immune system to an implanted device. Adsorption of biomolecules may cause creating the biofilms of bacteria or blood components which can lead to infection or clogging. In this regard, biomolecule adsorption is a necessary aspect in determining the long-term efficacy of lab-on-chip systems, implants, and medical devices that contact blood or other biological fluids (Finch, 2011).

The preparing of the microparticles with controlled shapes, compartments, and size distributions are an essential parameter which is considered during design and application of the different equipment including fixed-bed reactors, absorption columns, fluidized beds and cyclone separators for wastewater treatment. Attempts to achieve such features not only could provide a high adsorption capacity but also can enhance process efficiency and reliability. Despite the conventional procedures that have limitations to achieve precise control of size distribution, the microfluidic-based platforms with its intrinsic properties in a high degree of product control are an alternative technology toward the production of advanced emulsion-based templates and/or microparticles. Microfluidic systems are desirable for the production of particles with the narrow size distribution. For the production of particles with analogous properties, they should experience similar conditions in the reactor. This situation is difficult to achieve in batch systems owing to mixing limitations related to their large size. The formation of such particles is feasible in the microscale using continuous flow in the microfluidic devices (Fernandez Rivas & Kuhn, 2016). The experiments for adsorption of dyes pollutants using monodispersed chitosan microparticles synthesized on a microfluidic platform shows high removal efficiency (Dong et al., 2015). The high performance of microfluidic-synthesized technique can be a promising potential for future wastewater treatment processes.

6.7 Monolithic structures

Monolithic structures have some major benefits like mass transfer and/or heat transfer improvement as compared to the conventional shaped adsorbents like beads, granules or pellets.

These monolithic structures can be prepared via direct extrusion of the adsorbent or attaching an adsorbent layer to a support monolith by means of wash-coating, dip-coating, or crystal growth (Couck et al., 2018). However, these methods have some limitations such as flexibility. During usage of the support monolith with an adsorbent coating, a large part of the monolithic body does not contribute to the adsorption process. Under this condition, the volumetric yield of the adsorbent bed decreases. Furthermore, extruded monoliths which made from the active material itself, but the shape of the resulting monolith is determined by the extrusion die, impose a high cost on the system. This undesirable item limits the possibilities to study some effect of structural properties such as the wall thickness and channel width on mass and heat transfer in a systematic procedure. The 3D printing technology can be considered as a promising alternative to removing existence limitations (Fig. 6.11).

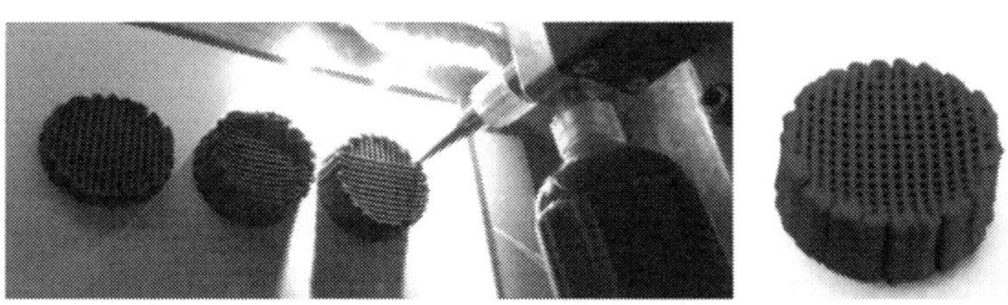

Figure 6.11 3D printed UTSA-16 monoliths for the adsorption of different gases (Grande et al., 2020).

This technology has a much higher level in freedom of operation, while it includes several printing methods like the fused deposition modeling (FDM), direct ink writing (DIW), stereolithography (SLA), and selective laser sintering (SLS).

The 3D printing technique is efficient for preparing structured adsorbents such as SAPO-34 zeolite. The SAPO-34 zeolite with honeycomb-like monolithic structures has been used for CO_2 adsorption from a CO_2/N_2 mixture which breakthrough experiments results indicate that good performance can be achieved, with very high selectivity for CO_2 (Couck et al., 2018). The performance of the fabricated monoliths for VOCs adsorption process is presented in Table 6.5 which confirms that the 3D printing technique renders perspectives to develop structured adsorbents.

Table 6.5 The performance of various fabricated monoliths for removal of VOCs (Hosseini, Moghaddas, Soltani, & Kheawhom, 2020).

Catalyst	Support	Application	Remarkable results
MnCu	FeCr alloy metallic monoliths	Ethyl acetate toluene combustion	Good activity ethyl acetate and toluene with detection CO_2 and H_2O
$MnOx/Al_2O_3$	FeCr alloy metallic monoliths	Combustion of volatile organic compounds	30 wt% loading indicated higher performance for ethanol, ethyl acetate, toluene
M-clay	Extruded monolith	Oxidation air pollutants	OH groups of bronsted acids and coordinately unsaturated cations, such as Al^{3+}, Fe^{3+} and Fe^{2+}. Lewis acids contribute to catalytic performance of air oxidation process.
Mn, Pt, Mn/Pt	Ceramic monoliths (natural clays and gibbsite)	Removal of $CHCl_3$ MEK Toluene Xylene	Conversion% Mn Pt Pt/Mn $CHCl_3$ 82 88 75 MEK 85 9 98 Toluene 2 55 79 Xylenes 23 64 90
Manganese	Washcoated pillared clay honeycomb monoliths	Oxidation of propane and acetone	The pillared clay allows incorporating higher amount of well distributed MnO_2

Table 6.5 The performance of various fabricated monoliths for removal of VOCs (Hosseini, Moghaddas, Soltani, & Kheawhom, 2020).—cont'd

Catalyst	Support	Application	Remarkable results	
Phenolic resins and cellulosic binders	Casting method	Toluene Benzene Xylene Hexane	Monolith activate carbon 18.3 g/100 g 11.1 14.7 13.0	 22.5 g/100 g 31.1 27.5 25.0
$CuO-CeO_2$	$SiO_2-Al_2O_3$ modified Glass-fiber Honeycomb	Oxidation Ethyl acetate Isopropanol Toluene	CeO_2 increase adsorption capacity of ethyl acetate, isopropanol and low activity in the oxidation of toluene	

6.8 Economic evaluation/justification

Although new designs in the field of equipment geometry have been able to create high efficiencies for sorption processes, in some cases the costs of construction and maintenance have limited their ability for large-scale application. The challenges of cost, manufacturing tolerances and channel plugging still remain an important issue for the use of microfluidic devices. The inability to induce turbulent flow, a high ratio of surface area to volume, and high shear rates are the characterization of the microfluidic systems. For very small length scales, the flow of a fluid is completely laminar and it is virtually impossible to induce turbulence, hence the parallel streams will mix only by diffusion. Therefore, this is a slower process than the turbulent mixing which can be achieved by a macroscale system and as a result, the creation of mixers is a great challenge for microfluidic devices. Microfluidic-synthesized materials face limitations for large-scale processes which is related to the high costs and limited production capacity associated with the microfluidic procedure.

6.9 Future opportunities and challenges

Despite the high performance of some conventional equipment, they suffer from inherent limitations (e.g., mass transfer rate) and its application in industrial-scale has faced problems. Some novel designs such as parabolic collectors and NETmix can be considered as alternative solutions. These devices operate in similar conventional reactor volumes, at high flow rates which can provide high efficiency.

The large-scale sorption processes using MCs has yet to be reported owing to the difficulty in the scale-up step. The scale-up of the MCs is also imposed relatively higher cost than the other technologies. The equal fluid control requires high precise of the fabrication to monitor the uniform distribution of solvent and gas flow rate in the microchannel. The conventional production methods such as laminate and molding, which are failure to address the commercialization issues for microchannel devices due to the fabrication speed and cost concern. Nowadays, the feasibility of using 3D printing method has been evaluated to replace the conventional production methods for commercial manufacturing. The fabrication of complicated geometry for microchannel reactor using metal material can be realized using the advanced 3D printing technology. However, high-pressure drop and low tolerance to particulate matter are known as the main disadvantages of this technology. Furthermore, the maintenance of such equipment is also relatively high since the suspended particulate matter is difficult to be removed or cleaned. Recent improvements in microdevices and their scale-up/integration technologies are a great potential for future large-scale. Nevertheless, fundamental research is a crucial aspect which should be considered to fill the gap by exploring the possibilities and potential benefits of using microfluidic devices.

6.10 Conclusion

The traditional devices which are used in the chemical industries suffer from some drawbacks such as low transfer efficiency. Therefore, new technologies should be applied to design efficient equipment with high mass transfer performance, energy-saving, and smaller volume. The current chapter proposed compact devices as a result of the modifications in the geometry to reduce the size and volume, while most of them have low installation, maintenance and operation costs. The findings in this chapter can serve as guidance in the design of intensified equipment and/or processes with high operational performance.

References

Ahmed, S. M. R., Phan, A. N., & Harvey, A. P. (2018). Mass transfer enhancement as a function of oscillatory baffled reactor design. *Chemical Engineering and Processing-Process Intensification, 130*, 229–239.

Assmann, N., Kaiser, S., & Rudolf von Rohr, P. (2012). Supercritical extraction of vanillin in a microfluidic device. *The Journal of Supercritical Fluids, 67*, 149–154.

Awasthi, A., & Datta, D. (2019). Application of Amberlite XAD-7HP resin impregnated with Aliquat 336 for the removal of Reactive Blue-13 dye: Batch and fixed-bed column studies. *Journal of Environmental Chemical Engineering, 7*, 103502.

Babar, M., Bustam, M. A., Ali, A., Maulud, A. S., Shafiq, U., Shariff, A. M., et al. (2019). Efficient CO_2 capture using NH_2—MIL—101/CA composite cryogenic packed bed column. *Cryogenics, 101*, 79—88.

Baldea, M., & Edgar, T. F. (2018). Dynamic process intensification. *Current Opinion in Chemical Engineering, 22*, 48—53.

Chemtech, S. (1997). Mixing and reaction technology. *Sulzer Brochure, 23*, 06.40-100.

Chung, C. I. (2019). *Extrusion of polymers: Theory & practice*. Carl Hanser Verlag GmbH Co KG.

Couck, S., Cousin-Saint-Remi, J., Van der Perre, S., Baron, G. V., Minas, C., Ruch, P., et al. (2018). 3D-printed SAPO-34 monoliths for gas separation. *Microporous and Mesoporous Materials, 255*, 185—191.

Díez, A. M., Moreira, F. C., Marinho, B. A., Espíndola, J. C. A., Paulista, L. O., Sanromán, M. A., et al. (2018). A step forward in heterogeneous photocatalysis: Process intensification by using a static mixer as catalyst support. *Chemical Engineering Journal, 343*, 597—606.

Dimian, A. C., Bildea, C. S., & Kiss, A. A. (2014). Process intensification. In *Computer aided chemical engineering*. Elsevier.

Domínguez, C. C., & Gamse, T. (2016). Process intensification by the use of micro devices for liquid fractionation with supercritical carbon dioxide. *Chemical Engineering Research and Design, 108*, 139—145.

Dong, Z., Xu, H., Bai, Z., Wang, H., Zhang, L., Luo, X., et al. (2015). Microfluidic synthesis of high-performance monodispersed chitosan microparticles for methyl orange adsorption. *RSC Advances, 5*, 78352—78360.

Fakhfakh, N., Dammak, N., & Benzina, M. (2018). Breakthrough modeling and experimental design for o-xylene dynamic adsorption onto clay material. *Environmental Science and Pollution Research, 25*, 18263—18277.

Fernandez Rivas, D., & Kuhn, S. (2016). Synergy of microfluidics and ultrasound: Process intensification challenges and opportunities. *Topics in Current Chemistry, 374*, 225—254.

Finch, C. (2011). *Modeling transport and protein adsorption in microfluidic systems*.

Firuzi, S., & Sadeghi, R. (2018). Simulation of carbon dioxide absorption process by aqueous monoethanolamine in a microchannel in annular flow pattern. *Microfluidics and Nanofluidics, 22*, 109.

Grande, Carlos, A., Blom, R., Middelkoop, V., Matras, D., Vamvakeros, A., Jacques, S. D. M., et al. (2020). Multiscale investigation of adsorption properties of novel 3D printed UTSA-16 structures. *Chemical Engineering Journal, 402*, 126166.

Graça, C. A. L., Lima, R. B., Pereira, M. F. R., Silva, A. M., & Ferreira, A. (2020). Intensification of the ozone-water mass transfer in an oscillatory flow reactor with innovative design of periodic constrictions: Optimization and application in ozonation water treatment. *Chemical Engineering Journal, 389*, 124412.

Hosseini, S., Moghaddas, H., Soltani, S. M., & Kheawhom, S. (2020). Technological applications of honeycomb monoliths in environmental processes: A review. *Process Safety and Environmental Protection, 133*, 286—300.

Huang, Y., Nie, Z., Yuan, J., Murray, A., Li, Y., Woods-Chabane, G., et al. (2019). Experimental validation of a test to estimate the remaining adsorption capacity of granular activated carbon for taste and odour compounds. *Environmental Sciences: Water Research and Technology, 5*, 609—617.

Hussein, M. S., & Ahmed, M. J. (2016). Fixed bed and batch adsorption of benzene and toluene from aromatic hydrocarbons on 5A molecular sieve zeolite. *Materials Chemistry and Physics, 181*, 512—517.

Iliuta, I., & Larachi, F. (2018). CO_2 and H_2S absorption by MEA solution in packed-bed columns under inclined and heaving motion conditions-hydrodynamics and reactions performance for marine applications. *International Journal of Greenhouse Gas Control, 79*, 1−13.
Jaberi, H., Mosleh, S., & Dashtian, K. (2020). Development of cigarette carbonaceous hydrochar/ZIF-67-based fluids for CO_2 capture from a gas stream in a packed column: Mass-transfer performance evaluation. *Energy and Fuels, 34*, 7295−7306.
Jaberi, H., Mosleh, S., Dashtian, K., & Salehi, Z. (2020). *Fluid based cigarette carbonaceous hydrochar supported ZIF-8 MOF for CO2 capture process: The engineering parameters determination for the packed bed column design. 153*, p. 108001. Chemical Engineering and Processing-Process Intensification.
Jähnisch, K., Baerns, M., Hessel, V., Ehrfeld, W., Haverkamp, V., Löwe, H., et al. (2000). Direct fluorination of toluene using elemental fluorine in gas/liquid microreactors. *Journal of Fluorine Chemistry, 105*, 117−128.
Jiang, X., Qing-Da, A., Xiao, Z.-Y., Zhai, S.-R., & Shi, Z. (2019). Versatile core/shell-like alginate@ polyethylenimine composites for efficient removal of multiple heavy metal ions (Pb^{2+}, Cu^{2+}, CrO_4^{2-}): Batch and fixed-bed studies. *Materials Research Bulletin, 118*, 110526.
Kang, M.-K., Kim, I.-D., Kim, B.-J., Kang, J.-S., & Oh, K.-J. (2015). Characteristics of absorption/regeneration of AMP/MEA using hybrid packing for the improvement of CO_2 absorption performance. *Industrial and Engineering Chemistry Research, 54*, 5853−5861.
Kurade, M. B., Waghmode, T. R., Xiong, J.-Q., Govindwar, S. P., & Jeon, B.-H. (2019). Decolorization of textile industry effluent using immobilized consortium cells in upflow fixed bed reactor. *Journal of Cleaner Production, 213*, 884−891.
Li, T., Fan, J., & Sun, T. (2020). Effective removal of methylene blue dye from water with nanocomposite ceramsites in a fixed-bed column. *Environmental Technology*, 1−13.
Li, M., & Li, D. (2017). Separation of Janus droplets and oil droplets in microchannels by wall-induced dielectrophoresis. *Journal of Chromatography A, 1501*, 151−160.
Lima, M. J., Silva, C. G., Adrián, M. T. S., Lopes, J. C. B., Dias, M. M., & Faria, J. L. (2017). Homogeneous and heterogeneous photo-Fenton degradation of antibiotics using an innovative static mixer photoreactor. *Chemical Engineering Journal, 310*, 342−351.
Lima, M. J., Silva, A. M. T., Silva, C. G., Faria, J. L., Lopes, J. C. B., & Dias, M. M. (2016). An innovative static mixer photoreactor: Proof of concept. *Chemical Engineering Journal, 287*, 419−424.
Lucas, M. S., Reis, N. M., & Li Puma, G. (2016). Intensification of ozonation processes in a novel, compact, multi-orifice oscillatory baffled column. *Chemical Engineering Journal, 296*, 335−339.
Mahdi, Z., Yu, Q. J., & Hanandeh, AEl (2019). Competitive adsorption of heavy metal ions (Pb^{2+}, Cu^{2+}, and Ni^{2+}) onto date seed biochar: Batch and fixed bed experiments. *Separation Science and Technology, 54*, 888−901.
Mi, Y., Yao, C., & Chen, G. (2018). Ethylene/ethane separation and mass transfer characteristics through absorption by $AgNO_3$ solutions in microchannels. *Chemical Engineering and Processing-Process Intensification, 130*, 110−118.
Mohtadi, M., James, B. R., & Davis, A. P. (2020). Activated carbon column adsorption of compounds that mimic urban stormwater dissolved organic nitrogen. *Water Environment Research.*
Moreno-Pérez, J., Bonilla-Petriciolet, A., Rojas-Mayorga, C. K., Mendoza-Castillo, D. I., Mascia, M., & Errico, M. (2016). Adsorption of zinc ions on bone char using helical coil-packed bed columns and its mass transfer modeling. *Desalination and Water Treatment, 57*, 24200−24209.
de Oliveira, L. H., Meneguin, J. G., Pereira, M. V., da Silva, E. A., Grava, W. M., et al. (2019). H_2S adsorption on NaY zeolite. *Microporous and Mesoporous Materials, 284*, 247−257.

Patel, H. (2019). Fixed-bed column adsorption study: A comprehensive review. *Applied Water Science, 9*, 45.

Pereira, F. M., Sousa, D. Z., Alves, M. M., Mackley, M. R., & Reis, N. M. (2014). CO_2 dissolution and design aspects of a multiorifice oscillatory baffled column. *Industrial and Engineering Chemistry Research, 53*, 17303–17316.

Rabha, S., Schubert, M., Grugel, F., Banowski, M., & Hampel, U. (2015). Visualization and quantitative analysis of dispersive mixing by a helical static mixer in upward co-current gas–liquid flow. *Chemical Engineering Journal, 262*, 527–540.

Rahmani, F., Mowla, D., Karimi, G., Ali, G., & Rahmatmand, B. (2015). SO_2 removal from simulated flue gas using various aqueous solutions: Absorption equilibria and operational data in a packed column. *Separation and Purification Technology, 153*, 162–169.

Rho, H. S., Hanke, A. T., Ottens, M., & Gardeniers, H. (2017). A microfluidic device for the batch adsorption of a protein on adsorbent particles. *Analyst, 142*, 3656–3665.

Scheiff, F., Mendorf, M., Agar, D., Reis, N., & Mackley, M. (2011). The separation of immiscible liquid slugs within plastic microchannels using a metallic hydrophilic sidestream. *Lab on a Chip, 11*, 1022–1029.

Sounthararajah, D. P., Loganathan, P., Kandasamy, J., & Vigneswaran, S. (2016). Column studies on the removal of dissolved organic carbon, turbidity and heavy metals from stormwater using granular activated carbon. *Desalination and Water Treatment, 57*, 5045–5055.

Sui, H., Jiang, P., Li, X., Liu, J., Li, X., & Lin, H. (2019). Binary adsorption equilibrium and breakthrough of n-butyl acetate and p-xylene on granular activated carbon. *Industrial and Engineering Chemistry Research, 58*, 8279–8289.

Tay, W. H., Lau, K. K., Lai, L. S., Shariff, A. M., & Wang, T. (2019). Current development and challenges in the intensified absorption technology for natural gas purification at offshore condition. *Journal of Natural Gas Science and Engineering*, 102977.

Tonkovich, A. L., & Daymo, E. (2017). Process intensification. In F. A. Kulacki (Ed.), *Handbook of thermal science and engineering*. Springer International Publishing: Cham.

Valencia, A., Ordonez, D., Wen, D., McKenna, A. M., Chang, N.-B., & Wanielista, M. P. (2020). *The interaction of dissolved organic nitrogen removal and microbial abundance in iron-filings based green environmental media for stormwater treatment* (p. 109815). Environmental Research.

Wu, J., Jia, L., Wu, L., Long, C., Deng, W., & Zhang, Q. (2016). Prediction of the breakthrough curves of VOC isothermal adsorption on hypercrosslinked polymeric adsorbents in a fixed bed. *RSC Advances, 6*, 28986–28993.

Xavier, A. L. P., Adarme, O. F. H., Furtado, L. M., Ferreira, G. M. D., da Silva, L. H. M., Gil, L. F., et al. (2018). Modeling adsorption of copper (II), cobalt (II) and nickel (II) metal ions from aqueous solution onto a new carboxylated sugarcane bagasse. Part II: Optimization of monocomponent fixed-bed column adsorption. *Journal of Colloid and Interface Science, 516*, 431–445.

Xiong, W.-L., Zhang, J., Jun-Xia, Y., & Chi, R. A. (2019). Competitive adsorption behavior and mechanism for Pb^{2+} selective removal from aqueous solution on phosphoric acid modified sugarcane bagasse fixed-bed column. *Process Safety and Environmental Protection, 124*, 75–83.

Yao, C., Zhu, K., Liu, Y., Liu, H., Jiao, F., & Chen, G. (2017). Intensified CO_2 absorption in a microchannel reactor under elevated pressures. *Chemical Engineering Journal, 319*, 179–190.

Yuan, J., Huang, Y., Nie, Z., & Hofmann, R. (2020). *'The effect of water temperature on the removal of 2-methylisoborneol and geosmin by preloaded granular activated carbon* (p. 116065). Water Research.

Zhang, D., Zhang, M., Ding, F., Liu, W., Zhang, L., & Cui, L. (2020). Efficient removal of formaldehyde by polyethyleneimine modified activated carbon in a fixed bed. *Environmental Science and Pollution Research*, 1–8.

Ziobrowski, Z., Krupiczka, R., & Rotkegel, A. (2016). Carbon dioxide absorption in a packed column using imidazolium based ionic liquids and MEA solution. *International Journal of Greenhouse Gas Control, 47*, 8–16.

CHAPTER SEVEN

Application of the supercritical fluids (SCFs) in the sorption processes

7.1 Introduction

All substances have a critical point in which the temperature and pressure conditions of each substance are different. When a compound is in conditions above its critical pressure and temperature point, it is said that the liquid has become "supercritical." As can be seen in Fig. 7.1, the gas–liquid equilibrium curve is interrupted at the critical point, building a coherence of physicochemical properties. The physicochemical properties of substances which is most often used as supercritical media are presented in Table 7.1.

Above the critical temperature, increasing pressure lead to the disappear phase boundary and formation of the supercritical fluid (SCF). It is shown in Fig. 7.1. After that, gas cannot be liquefied at any pressure. The critical point C is the end of the gas–liquid equilibrium curve on the temperature-pressure plane. The SCFs possess liquidlike densities, gaslike viscosities and diffusities intermediate to that of a liquid and a gas. Because of these two key properties, the SCFs are used as extractants. The low viscosity causes quick permeation of the solvent into the solid matrix, whereas, the relatively high density contributes to SCFs good solubilization (Berk, 2018).

This specific behavior of substances was first observed in 1822 by Charles Cagniard de La Tour in his renowned cannon barrel experiment. It was afterward determined as an SCF by Irish chemist Thomas Andrews. CO_2 (carbon dioxide) and water are the most widely used SCFs (Kohli, 2019; Weibel & Ober, 2003).

High-pressure technologies encompass supercritical fluids proposal the feasibility to gain new products with singular features which are environmentally supportable and friendly (Knez et al., 2019). Supercritical solvents have been increasingly interested in the last few decades. One reason is that their solvent properties change considerably with temperature and density. They are "flexible solvents" and for each goal, the optical properties can be adjusted. The enhanced heat and mass transfer is another advantage of

Intensification of Sorption Processes
ISBN: 978-0-12-821411-4
https://doi.org/10.1016/B978-0-12-821411-4.00001-8

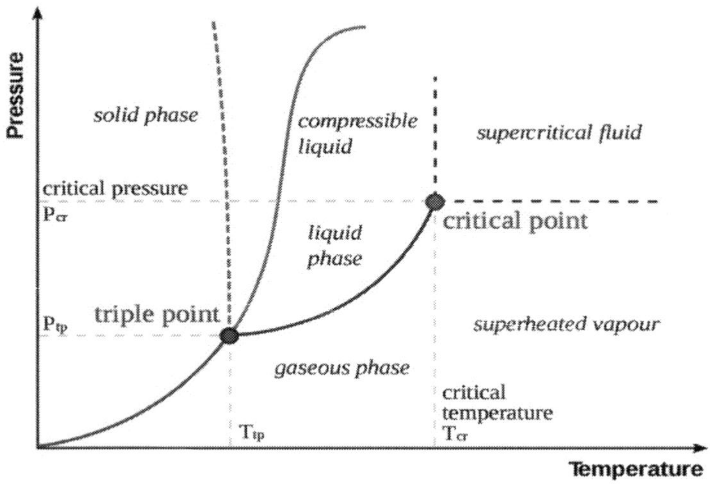

Figure 7.1 Phase diagram showing the supercritical region (Berk, 2018).

Table 7.1 Physicochemical properties of some SCFs (Knez, Pantić, Cor, Novak, & Hrncic, 2019).

Supercritical fluid	Name	T_c[a](°C)	p_c[b](bar)	μ[c] (Debye)
Ar	Argon	−122.5	48.6	0
CO_2	Carbon dioxide	31.1	73.8	0
H_2O	Water	374.0	220.6	1.85
SF_6	Sulfur hexafluoride	45.5	37.6	0
Xe	Xenon	16.6	58.3	0
CHF_3	Fluoroform	25.9	48.2	1.65
CH_2F_2	Difluoromethane	78.1	57.8	1.97
C_2H_6O	Dimethylether	126.9	54.0	1.30
C_3H_8	Propane	96.7	42.5	0.084

[a]critical temperature,
[b]critical pressure,
[c]electric dipole moment.

SCFs, which show their good solvent properties only in the supercritical state. Hence, the separation process can be simply attained after reaction or extraction by decreasing temperature and pressure (Mendes & Coelho, 2005).

Different supercritical solvents are used for practical purposes including carbon dioxide, methane, ethane, propane, butane, pentane, tetrafluoromethane, trifluoromethane, sulfur hexafluoride, dimethyl ether, acetone, methanol, ammonia, and water. carbon dioxide is the most employed supercritical solvent due to its special characteristics and physicochemical properties, for example, no flammability, relatively inexpensive, nontoxicity, higher mass transfer rate, and adjustable extraction power for organic compounds based on the density (Sun, 2002). Much activity exists in the extraction of natural products by using SCFs such as pure carbon dioxide and carbon dioxide with moderators. Some processes such as the decaffeination of coffee and tea, the extraction of hops, spices, and drugs are among important industrialized applications. Applications including the extraction of rapeseed or the separation of glycerides or butter constituents are being developed. The other promising application of supercritical fluids as solvents is dyeing processes (e.g., of polyester with dispersion dyes), the precipitation of solids (e.g., explosives, coatings), polymer chemistry (e.g., for the extraction and separation of oligomers), chemical reactions (e.g., treatment of wastes by supercritical steam) and many other fields (Mendes & Coelho, 2005).

As a first approximation, the solvent strength of an SCF can be attributed to the solution density. One of the principal preponderances of SCF reaction media is that the density can be changed continuously from liquidlike to gaslike values by changing either the temperature or the pressure. Table 7.2 shows the lists of the pure-component critical properties for selected solvents that have been considered as SCF reaction media. As shown, these solvents cover a relatively broad range of critical temperatures and pressures (Sun, 2002).

Table 7.2 The critical properties of selected components.

Component	Critical temperature (K)	Critical pressure (MPa)	Critical density (kg/m^3)
Carbon dioxide	304.13	7.38	467
Ethanol	513.90	6.14	278
Ethane	305.46	4.87	206
Methane	190.56	4.59	162
Propane	309.52	4.25	228
Ammonia	406	11.28	230
Acetone	508.10	4.70	278
Ethylene	282.40	5.04	215
Water	647.30	22.12	348
Propylene	365	4.62	230

From: Sun, Y.-P. (2002). Supercritical fluid technology in materials science and engineering: Syntheses: Properties, and applications. CRC Press.

7.2 Some advantages of supercritical fluids

The superiority of using a SCF can be summarized as follows: SCFs have similar solvating powers to liquid organic solvents, but their better diffusivities, lower viscosity, and lower surface tension make them more efficient in many processes. because of their adjustable density (Cansell, Rey, & Beslin, 1998) with pressure, separation of substances from solvents is easy. Have this capability to add modifiers to an SCF, for variation the polarity or gives them more adoptive separation power. SCFs are inexpensive, safe to use and have a low operating cost (Kazarian, 2000). SCFs as compared to volatile organic compounds (VOCs) and ozone-depleting substances (ODSs) are more environment-friendly because of their less residual waste in the process (Goodship & Ogur, 2005).

7.3 Supercritical fluids sorption mechanism

It is recognized that "adsorption" is a surface-based process while "absorption" involves the whole volume of the material; the term "sorption" encompasses both processes. Therefore, the sorption of gas or organic vapor on a substance generally includes surface adsorption and filling of substantial volume simultaneously. The understanding of the solute retention mechanism in the SCF relies on determining the complex interaction between multiple chemical processes. To investigate the actual sorption mechanism, methods should be used that can examine sorbed elements and determine the nature of species and their types of binding. This aim can be attained using a multidisciplinary approach, with various methods including surface methods, bulk spectroscopy, and microscopic methods that provide supplementary results (Fédoroff & Lefèvre, 2006). For example, Table 7.3 shows the main sorption mechanisms for some elements on hydroxyapatites which were investigated by Fedoroff Monteil-Rivera et al. (2000). The significant characteristic is that the sorption mechanisms are extremely related to the element. In concerning hydroxides, the general primaries of surface complexity can be applied, but it is difficult to determine the locations and complexes of actual sorption and thermodynamic constants. SCF chromatography is more practical in virtue of its solvent compression properties and higher working pressures. In supercritical liquids, there is no surface tension due to the lack of a boundary between the gas and liquid phases.

Table 7.3 Main sorption mechanisms of some elements on hydroxyapatites.

Element	Sorption mechanisms
Cd^{2+}	Exchange with Ca^{2+} in the crystal framework in two sites, with diffusion in three superficial unit cells. Equilibrium not achieved within 10 d at 75°C
Pb^{2+}	Formation of new solid phases + slight incorporation into apatite framework
UO_2^{2+}	Formation of new amorphous solid phase
Eu^{3+}	Superficial sorption (site 1) + incorporation in the apatite framework (site 2)
SeO_3^{2-}	Exchange with PO_4 groups of one superficial unit cell

By changing the temperature, the properties of the fluid can be adjusted to behave like a gas or liquid. The solubility of a SCF is directly related to its density. One of the most important properties of SCF is solubility (Yonker & Smith, 1990).

Common processes in sorption are encompassing the regeneration of porous beds, preparative scale supercritical chromatography, simulated moving beds, thermal swing schemes, and adsorption/desorption cycles. Although initial applications of supercritical fluids in this field were on the regeneration of porous beds, lately, more focus is on fractionation reflects which can be met the real potential of the technique. Some examples are presented in Table 7.4 (Sunol & Sunol, 2001).

7.4 Separations using supercritical fluids

This section is an overview of the principles of using supercritical fluid (SCF) for separation. Extraction and chromatographic are the separation methods which are the oldest and the most powerful current respectively. SCF chromatography (SFC) and supercritical fluid extraction (SCFE) have appeared as favored technologies for rapid, contamination free extraction, and treatment in the pharmaceutical and food industries. According to studies by Agrawal et al., supercritical liquid chromatography is faster and found to be more environmentally friendly than natural phase liquid chromatography in separation. Due to emphasis on utilization of environmental compatibility selection, the use of supercritical liquid chromatography is recommended to determine chiral purity in different quality control laboratories and baseline separation (Agrawal, Belemkar, & Bonde, 2020).

Table 7.4 Sorption applications of supercritical fluids.

Purpose	Supercritical medium	Substrate
Environmental		
Soil remediation	CO_2 + entertainers	Soil
Regeneration of adsorbents	CO_2	Polychlorinated biphenyls from adsorbents
Petrochemical		
Regeneration of adsorbents	CO_2	Benzene, toluene, ethyl acetate from activated carbon
Regeneration of catalysts	CO_2, hexane, ethylbenzene, THF, benzene	Hydrocarbon, coke removal from catalysts
Chromatographic fractionation	CO_2	Coal tar, mineral oil distillate
Essential oils		
Chromatographic fractionation	CO_2 + alcohol	Black pepper, clove extract
Polymer processing		
Chromatographic fractionation	CO_2, n-pentane + methanol	Styrene oligomer
Biochemical		
Chromatographic fractionation	CO_2 + alcohol	Prostaglandins
Chromatographic fractionation	CO_2 + alcohol	Vitamins, polyunsaturated fatty acids
Chromatographic chiral separation	CO_2	Optical isomers

SCFs are effective at much lower temperatures than distillation, and the application of SCFs in separation prevents degradation and the break-down of heat-sensitive compounds. The sensitivity of responses to process variables, energy savings, promise of complete and versatile regeneration of solvents, promoted solute volatilities, solvent selectivity, favorable transport properties for solvents, and state governed effectiveness of solvents which enables the use of low cost, nontoxic, environmentally acceptable solvents are the attractiveness point of supercritical extraction processes. The impact of the inherent attributes of supercritical fluids on separations is summarized in Table 7.5 (Sunol & Sunol, 2001).

Many attempts have been made to classify separation technologies. Supercritical fluids can apply to either intraphase or interphase separations. Among the technologies, intraphase is more common due to the ease and flexibility in which a new phase can be formed for the regeneration of the solvent. Whereas, interphase separations have been limited by material solubility and swelling problems, particularly with organic component-based membranes. This is caused by the increased solubility of these components in supercritical solvents. Generic steps involved in a typical chemical process have been presented in Fig. 7.2, with each separation subtask type indicated. Environmental processes are usually prevailed by feed preparation while biochemical processes apply all of the four subtasks. Splitting is

Table 7.5 The characteristics and challenges with supercritical separations.

Inherent characteristics of systems at supercritical Conditions	Resulting promise	Challenges to be met
Enhanced solubility	Effective at lower temperatures	Recovery of substances inadvertently extracted
Favorable transport properties	High mass transfer rates especially in porous media	
Inherent characteristics of systems at supercritical conditions	Resulting promise	Challenges to be met
The sensitivity of responses to process variables	Effective recovery of solvents, and innovative fractionation Possibilities	Control and optimization
High pressure	High capital cost and need for recovery of mechanical energy	
The dominance of physical characteristic on solvency over chemical	Expansion of solvent spate to environmentally benign, inexpensive, and nonflammable Solvents and their mixtures	

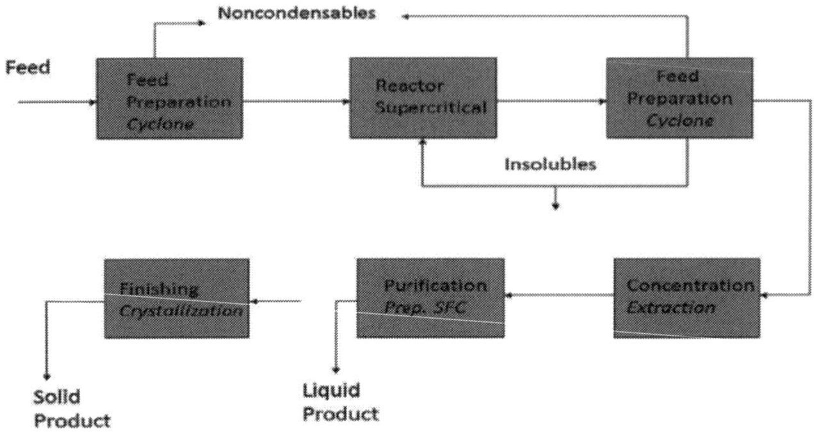

Figure 7.2 A generic process flowsheet with possible supercritical steps (Krukonis, 1988).

considered as a feed preparation duty while blending is considered as a finishing duty. The objective criterion is used for the separation may change according to the subtask. For example, in-feed preparation, when the usual economic objectives bound to technical feasibility limits become the purpose for the purification and finishing stages, technical feasibility such as removal of the fines is more appropriate. The feed preparation duty includes removal of insoluble components and noncondensable compounds at times. The technology selection and the scheduling of tasks are dependent on the concentration and particle size distribution. In common applications, these processes are naturally physical/mechanical. The unit designed for carrying out the feed preparation for the purpose or the function may be connected to another function within a single unit. For example, the fines, either nucleated by pressure and/or temperature disturbance, or particles suspended in the SCF can be easily separated by using cyclones Concentration (or isolation) duties are usually necessary with streams where the product is very dilute. Adsorption and extraction are typical technologies applied here. Energy considerations prevail in the technology selection subject to product contamination and stability limits. Thus, the basic technology is depended on the accessibility of the mass separating agent (e.g., adsorbent or solvent) and it's effective to recycle. A typical example of the isolation step is supercritical extraction of organics from aqueous streams.

Typical operations such as crystallization, drying, and blending are necessary to finish the product to its final form. An example of finishing is Particle nucleation from a SCF solution. Every supercritical separation technique is identified in this generic process.

7.4.1 Leaching: generic application

The leaching process is usually considered as extraction from a solid substrate that is not accurately correct, and that is commercially the most significant application of supercritical fluids. Fig. 7.3 shows a generic flowsheet, explains the basic concepts of processing. A heat-labile natural substrate usually comes in contact with a SCF in a semicontinuous manner. As an option, the SCF could be doped with an entrained. Due to the batch nature of the leaching stage, two or more parallel vessels filled with the substrate may be employed. The material leached is recovered and fractionated by pressure decrease, temperature turmoil, adsorption, membranes, or through the absorption process.

During the beginning phases of leaching, the solute loading is more rapid and may approach the equilibrium solubility limit, while at the latter phases due to kinetic and mass transfer constraints relatively lower concentration of the solute is occurred. Two popular commercial successes of this application are coffee decaffeination and hop extraction. Moreover, common applications include food processing, environmental and parts cleaning, bioactive compounds production, and fuel processing. Examples are presented in Table 7.6.

7.4.2 Extraction: generic applications

Liquid streams involving components with close boiling point and heat-sensitive can be divided into their components with the help of supercritical

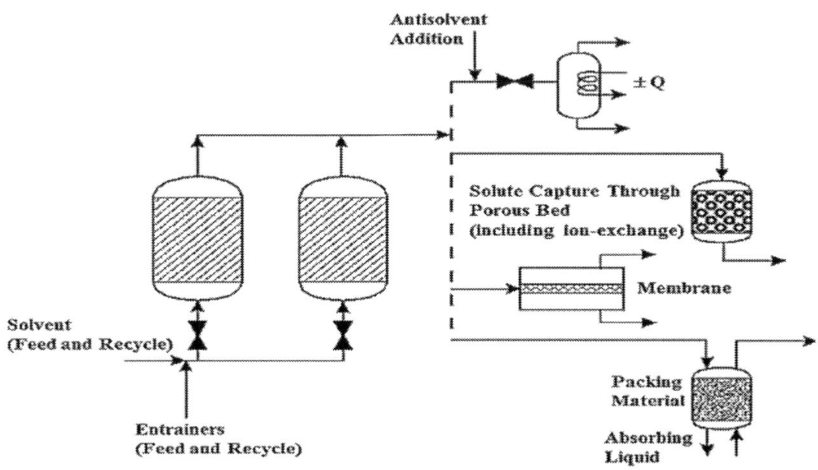

Figure 7.3 Supercritical leaching and options for supercritical solvent regeneration (Krukonis, 1988).

Table 7.6 Leaching applications of SCFs.

Application	Substrate	Supercritical solvent	Extract
Pharmaceutical	Cumin	Carbon dioxide	Antimicrobial oil
	Ginger	Carbon dioxide + methanol	Antipyretic
	Chamomile	Carbon dioxide	Antiinflammatory Antispasmodic
	Sunflower leaves	Carbon dioxide Carbon dioxide + water	Herbicide
	Rosehip, cherry, grape seed,rosemary, algae, microalgae	Carbon dioxide	Antioxidant
	Bamboo shavings	Carbon dioxide + cosolvent	Antifatigue
	Algae, microalgae	Carbon dioxide	Lipid, metabolites, pigments
Energy	Algae, microalgae	Carbon dioxide	Lipid, hydrocarbons, alkadines
	Coal, lignite	Toluene, water, acetone, methanol. THF, tetralin, Ethanol	Liquid products
Environmental	Soil	Carbon dioxide + modifier	Polycyclic aromatic hydrocarbons, pesticides
	Treated wood, soil, sand	Carbon dioxide + chelating agent	Heavy metals
	Solid fuels	Toluene, water, acetone, methanol. THF, tetralin, Ethanol	Sulfur, oxygen, nitrogen
	Soil	Nitrous oxide	Pesticides
	Soil	Water	Polycyclic aromatic hydrocarbons

fluids or valuable components of aqueous streams could be isolated and concentrated using supercritical solvents. Also, to promote selectivity, entertainers could be added to the solvent, especially in fractionation instances. The primitive cause of using entertainers for concentration objectives is to increase solubility. Membranes can also be used along with a supercritical fractionator. Typical applications of this process technology are given in Tables 7.7 and 7.8 while several review papers and texts provide useful design information.

Table 7.7 Extraction applications of supercritical fluids.

Purpose	Supercritical medium	Substrate
Food processing		
Food purification	CO_2	Oil purification cholesterol removal from milk fat
Flavor extraction	CO_2	Beer, wine and fruit flavor concentrates
Fractionation of food	CO_2	Oil, milk fat
Forest products		
Extraction of lignocelluloses' materials and bio mass	CO_2, SO_2, N_2O, H_2O, methylamine, ethanol, ethylene	Cellulose or lignocelluloses' materials from wood or wood pulp
Biochemical		
Medicinal compound	CO_2	Fermentation broth
Petrochemical		
Purifying liquid fuels	CO_2	Deasphaltation, demetallization
Environmental		
Wastewater extraction	CO_2+entrainer	Aqueous toxic wastes, phenol, metabolic wastes
Essential oils		
Purification of essential oils	CO_2	Terpenes from orange, citrus and bergamot peel oil

Table 7.8 Fractionation applications of supercritical fluids.

Purpose	Supercritical medium	Substrate
Essential oils		
Fractionation	CO_2	Glycerol, fatty acids, glycerides
Chemical		
Fractionation	CO_2	Xylenes, glycerides, alkanes, aromatics
Fractionation	CO_2 + entrainer	Methyl esters, fullerenes
Petrochemical		
Fractionation	Toluene, pentane, propane	Petroleum pitch, asphalt
Polymer processing		
CO_2	Polydimethylsiloxane, fluoropolymers, polypropylene	

The procedures practiced in the design of extraction systems are similar to the usual appreciation for liquid and gaseous systems and include Kremser, McCabe Thiele, Panchon Savarit, and more rigid rate-based as well as equilibrium based procedures. Due to the highly nonlinear nature of the physical and chemical characteristics in the critical region, great care should be taken in representing the properties and the solution algorithms (Krukonis, 1988).

Generally, extraction depends on this basic principle that the solubility of a given compound (solute) in a solvent varies with both temperature and pressure. The solubility of a solute in gas is generally ignorable at ambient conditions (25°C and 1 bar) and is related directly to the vapor pressure of the solute. In an SCF, however, solute solubility's up to 10 times greater than those predicted by ideal gas law. This is because in supercritical fluids the solubility of a solute does not only depend on pressure. Also, solute-solvent interaction, as well as the vapor pressure, affect the dissolution reaction. Therefore, volatile solids have a much larger solubility in SCF in general, than for example in ideal gases. To reduce the amount of solvent required, more increase of the solubility within an SCF extraction process may also be favorable. This may be attained by the addition of another component to the blend. This may be mentioned as a cosolvent or entrained. This material will generally have volatility intermediate to that of the SCF and the solute which enables controlling the reaction more precisely and selectively. However, the usage of excess materials may also alter the chemical nature of the SCF and the capacity to remove it after reactions must also be taken into account (Krukonis, 1988).

To make the fluid supercritical it is compressed to increased pressures above its critical pressure. The polymer is then subjected to the SCF and swells. While the void volume in the polymer is increased, the SCF can pierce deeply into the matrix and the impurities are dissolved by the supercritical fluid. As each volatile material within the feed matrix will then divide themselves within the supercritical phase, these are eliminated with the SCF during the extraction from the feed system. In the depressurization stage, the pressure is rapidly decreased and the SCF and impurities permeate out of the polymer. The SCF can then be eliminated by changing temperature and/or pressure, leaving ignorable SCF in the extracted material and the SCF can be recovered by recompression. This system is represented in Fig. 7.4. There are both advantages and disadvantages of using SCFs as compared to conventional liquid solvents for separations. Some of the advantages consist of:

- The dissolving capacity of the SCF is easily controlled and handled by pressure and/or temperature.
- No harmful residue is left due to using nontoxic solvents.
- Easy recovery and recycling of the SCF from the extract due to its volatility.
- The ability to sometimes gain separations that are not feasible by usual processes.
- Low temperatures can be used for the extraction of heat-sensitive materials.
- Promoted working pressure is required with associated costs and hazards.

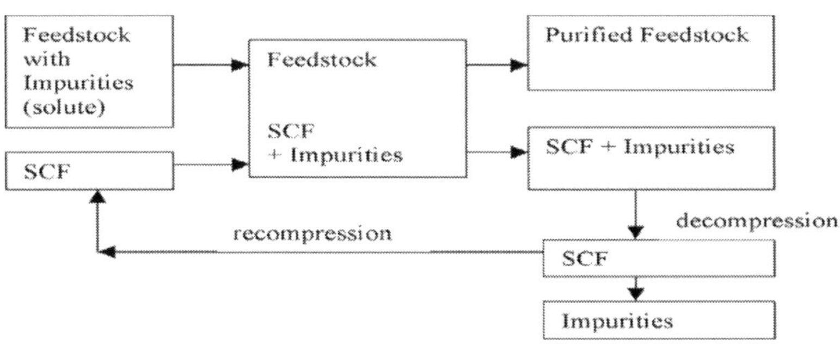

Figure 7.4 Basic principle of SCF extraction.

Whereas recycling and cost savings can be attained by recompression of the solvent system, equipment installation can be costly and complex. Therefore, a high capital cost is required for the equipment.

7.4.3 The supercritical fluid extraction process

In the various industrial segment, it have been trying to implement new technologies in the field of extraction. The technique of SCE is an example of technological advances in this area. The technology of extraction by using SCFs are a suitable alternative instead of traditional methods for the extraction of many compounds (Lindy, 2015).

7.4.4 Effective parameters in the extraction process

in using the SCE technology, there are some important parameters that will considerably affect the solvent potency and the mass transfer processes that occurred during the extraction. In the following, the most important parameters are defined (Sanchez-Camargo, Mendiola, Elena, & Herrero, 2014).

7.4.4.1 Raw materials

The raw material is defined as the sample to be extracted. For SCFE usage, either solid or liquid samples might be employed, although in each case the attention given as well as the instrumentation needed is little modified. in solid samples, the physical condition of the sample may have the most influence. Particle size and morphology will have the most affected on mass transfer by growing the surface contact although the moisture of the sample may also disturb the extraction process. Using the dried samples give better results, although exceptions exist. We should set the important parameters before the process. If the sample size is too small, building the channels inside the extraction cell is possible. To eliminate this problem, dispersing agents may be applied to generate homogeneous extractions.

In the liquid samples, countercurrent extractions are generally used to increase contact between the sample and the SCF. In these cases, the liquid sample is entered in the upper part of a packed extraction column and the supercritical CO_2 is entered from the bottom. By suitably selecting the entering point (height) as well as the inert column packing material that increases the surface contact, the mass transfer may be optimized.

7.4.4.2 Solubility (pressure and temperature)

Extraction pressure and temperature are perhaps the most affecting parameters in terms of the solubility of a material in the SCF. In the main, it can be said that the higher density of the SCF will be obtained by increasing pressure and will lead to an increase in the solubility of sample components. On the other side, the enhanced temperature will reduce the density (for a given pressure) although will also raise the transfer of solutes from the sample to the SCF due to the increase in their vapor pressure. so, select the temperature and pressure values have to be done carefully according to the aim of the process as well as the targeted compounds. To see the effect of temperature and pressure, the use of experimental is proposed.

7.4.4.3 Polarity/use of modifiers

CO_2 is the most used SCF nowadays, although its low polarity restricts some of its application to the extraction of low polar/lipophilic compounds. In order to increment the range of potential applications, a modifier might be used with supercritical CO_2. Typically, organic solvents like methanol or ethanol are used as modifiers, at a concentration below 10% related to the amount of CO_2 used for the extraction. In some applications, water has been used. in these conditions, the physical properties of the solvent mixture are more complex, more importantly, because the modifiers might not be in their supercritical mood; thus, different phases may be appearing during the extraction procedures. another modifier has been also employed to support the extraction of very low polarity components. Finally, it has to be attention that while using modifiers the possibility of getting solvent-free extracts is lost because these solvents are not gases at room state.

7.4.4.4 Solvent/feed ratio

The solvent/feed ratio is important in the supercritical process. after defining the pressure and temperature conditions, it is necessary to study the influence of the solvent/feed ratio or the effect of the CO_2 flow rate. This flow rate should be excellent value enough to maximize the extraction yield but also low enough to allow well contact with the sample to minimize the amount of CO_2 usage, and then, the operational costs. As it can be concluded, this parameter is very important while extracting liquids under countercurrent situations, as in those cases, the ratio will describe the contact time permissible between the sample and the supercritical CO_2.

7.4.4.5 Selecting solvent

Many parameters are the same as those used in more traditional extraction methods in selecting solvent of SCFE, for example, has good characteristics, is inert to the final material, can be readily separated, and is cost-effective to utilize. For SCF, that usually means that the critical parameters should be attainable without undue expense or complexity. That is why carbon dioxide is the most typically used SCF in SCFE, because of its low critical parameters, low cost, and low toxicity. SCW is one of the alternative solvents with high potential, due to its ability to dissolve organic matters while ignoring inorganic salts. The reverse effect is its uncritical state. This enables the same solvent to be used for both extractions. Some other SCFs like hexane and methylene chloride have been used commercially and investigated. For example, the usage of organic solvents in the petrochemical industry. However, these materials need to employ explosion-proof equipment, raising the cost of these processes. Other limits on solvent selection may also apply, for example, the limitation of the usage of one set of highly effective supercritical solvents (chlorofluorohydrocarbons, CFCs) because those affect the ozone layer. Carbon dioxide or water in the SCF form are the alternatives of organic solvents in both the food and medical industries. SCFs are used as safe substitute solvents in the food processing industry. They are also used in the rapid extraction of spirits and other components at room temperature. There are other fields of application including:

- Purification and fractionation of polymers such as the elimination of unchained monomers from polymers.
- Purification and separation of compounds such as oil and grease.
- Removing impurities in chemical materials.

Vacuum, steam stripping or solvent extraction are conventionally employed to purifying polymers. However, these methods are not always adequate as they may not able to reduce the residual contents to the required allowable levels. As an alternative, devolatilization with supercritical fluids can enhance the removal of impurities due to an increased thermodynamic driving force and enhanced molecular diffusivity. As a result, it is more effective. Other applications of this method are to remove residual solvents, additives, catalysts, and side reaction products, whereas, can have a harmful effect on the end-use properties. Since supercritical fluids are tunable, selective extractions can be carried out providing the SCF has adequate solubilizing power with the product to be purified (Krukonis, 1988).

7.4.5 Instrumentation for supercritical fluid extraction

Today, there is a large range of commercial instruments from Laboratory-scale to industrial scales to perform SCFE. However, it is common to discover applications based on in-house built equipment. The basic tools

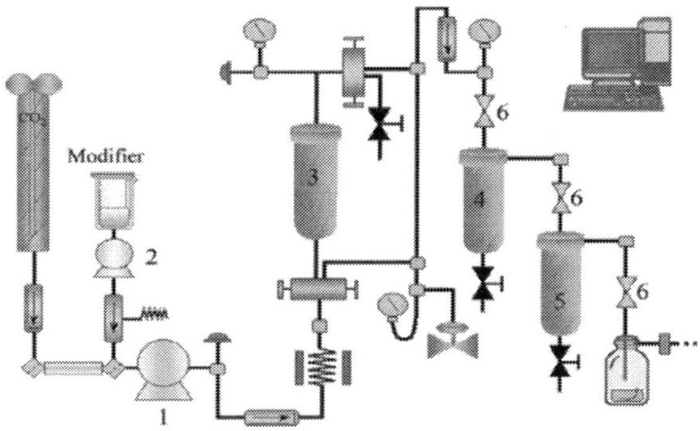

Figure 7.5 Essential instrumentation required for a supercritical fluid extraction equipment. *CV*, collection vessel; *M*, modifier; *S1*, separator 1; *S2*, separator 2 (Herrero, Cifuentes, & Ibañez, 2006).

needed to build an SCFE instrument will little vary depending on the application, solids or liquids extraction. In Fig. 7.5, the basic components of an SCFE extractor are shown.

The first section of each extractor is dedicated to the extraction process itself, composed of a temperature-controlled extraction chamber or column that can tolerate the high pressures needed to perform the extractions, a CO_2 pump and a modifier pump. In the case of liquid samples, the extraction column is often armed with different ports for the presentation of the sample at variable heights. In the second case, another extra pump is needed to enter the sample into the system. The CO_2 pump is the one ambiance the pressure inside the extraction chamber so that the SCF-CO_2 is often under the desired conditions, that is hold by using a back pressure regulator. The next part of the extractor is centralized on extract recovery. It may be built by a collection cell or by many fractionation vessels for performing depressurization (Herrero et al., 2006; Sanchez-Camargo et al., 2014).

7.5 Fundamentals for adsorption from supercritical phases

SCF can be considered as a mixture of liquid and gas; the supercritical gas has typical of the supercritical fluid with less density. As can be seen in Fig. 7.6, the radial dispensation function $g(r)$ of the SCF has abroad the second peak around $r = 0{\cdot}7$ nm. The radial dispensation function of liquid has

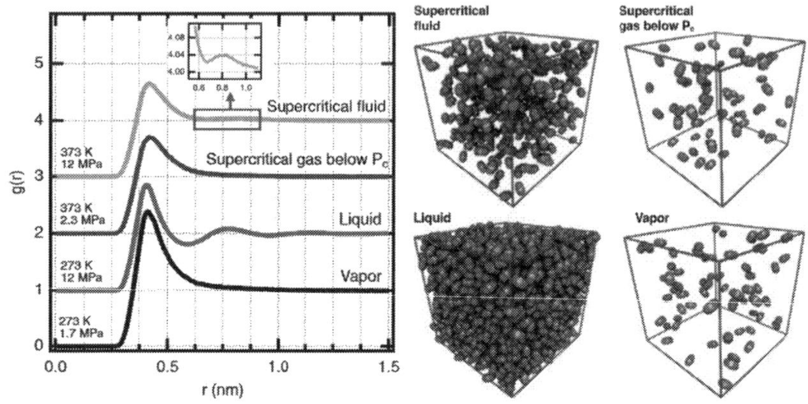

Figure 7.6 Radial dispensation functions and snapshots obtained from computer simulation for CO_2 in different phases: *SCF*, supercritical gas, liquid, and vapor (Vallejos-Burgos, Ohba, & Kaneko, 2019).

a clear second peak demonstrating, together with the intense first peak. Therefore, the supercritical gas below P_c has a short range order distinct from the supercritical liquid.

Adsorption is a process that molecules from a fluid phase connect themselves to the surface of the nearest phase. The surface of this phase comprises atoms or practical assembles as a segment of surface-connected molecules. Residual bulk and surface of the phase are the so-called adsorbent. The adsorbent surface atoms are located at the distance of the interacting forces between the bulk adsorbent phase and the molecules of the fluid phase (from which adsorption occurs). As can be seen in Fig. 7.7, adsorbent and adsorbed compounds (the adsorption) form the adsorbate. Interaction forces between bulk and surface molecules of the adsorbent define, across other features, the stability of the adsorbent in equilibrium with a SCF, which approximately functions at the surface as a solvent for the molecules. Interaction forces between bulk and surface molecules of the adsorbent define, across other features, the stability of the adsorbent in equilibrium with a SCF, which approximately functions at the surface as a solvent for the molecules. Interaction forces between molecules of the fluid phase and surface molecules define the adsorption equilibrium of molecules from the fluid phase. The interaction forces may be of physical essence with long ranging forces and no saturation, or of chemical essence with unique interactions and saturation after all surface molecules have reacted with compounds from the fluid phase. Fluid mixtures that contact with an adsorbent, type and size of the

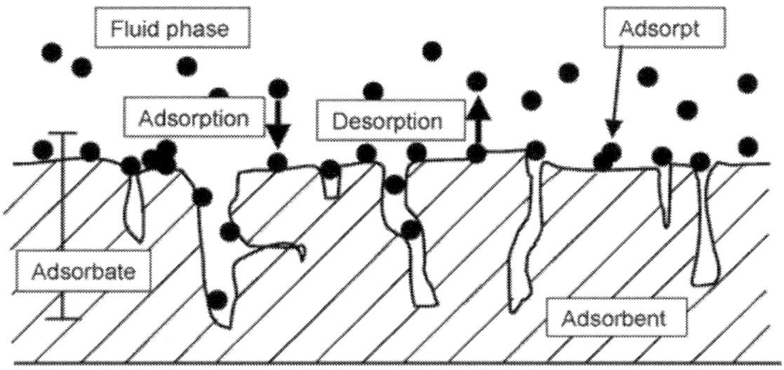

Figure 7.7 Adsorption at the surface of a solid (Brunner & Johannsen, 2006).

interaction forces between the surface molecules and the different molecules of the bulk fluid phase conduce to an adsorptive surface layer, which is different in composition from the fluid bulk phase. This diversity may form the foundation of separation processes and chemical reactions.

7.5.1 Adsorption isotherms for pure components

Adsorption isotherms for pure components used to describe the adsorption of single components at a constant temperature in most cases, simple and superior semiexperimental equations are used, because thermodynamic description of the adsorbent surface is not plausible for functional targets. The simplest form of an adsorption isotherm is a linear equation. In this case the loading of the adsorbate q_i is proportional to the concentration c_i of component i in the fluid phase (Eq. 7.1) with a constant adsorption coefficient K_i^H:

$$q_i = K_i^H c_i \cdot \tag{7.1}$$

However, for the most part, adsorption isotherms will be linear only in small ranges, at low sufficient concentrations the linear adsorption isotherm can be applied. This linear relation is widely used in analytical chromatography (Brunner & Johannsen, 2006).

About nonlinear behavior of adsorption equilibrium, several renowned equations have been expanded, for instance, the BET equation, the Langmuir equation, and the Freundlich equation. These extensively used equations are restricted to model isotherms with a single curvature. Isotherms that displaying both curvatures cannot be described (Vallejos-Burgos et al., 2019).

7.5.2 Multicomponent adsorption

Adsorption of more than one component on an adsorbent hint at interaction of these components resulting in different adsorption equilibrium in contrast to those of the pure components. Multi-ingredient adsorption can be a model for deriving the multi-component equilibrium from the one-component equilibrium of the pure substances (e.g., the Ideal Adsorbed Solution Theory (IAST)). IAST was suggested by Myers and Prausnitz (1965) for explaining and anticipating the adsorption equilibrium of mixtures from the adsorption of the pure components. The adsorption modeling of single components is performed by single adsorption isotherms. The adsorbent formed by the adsorbent components which is regarded as an ideal solution. The interplay between the components is ignored while the adsorption of the components is limited. The resulting equation accomplishes the Gibb–Duhem equation. As given by the following equation The composition of the adsorbate is formulated in mole fractions:

$$z_i = \frac{n_i^{\mathrm{ad}}}{n^{\mathrm{ad}}}; \quad \sum_i z_i = 1 \cdot \tag{7.2}$$

Where z_i is the molar fraction in adsorbate, n_i^{ad} is molar quantity in component i and ad is related to adsorbate. Fugacity of a component in the adsorbate f_i^{ad} is defined similarity to the fugacity of gases (Eq. 7.3).

$$d\mu_i^{\mathrm{ad}} = RT\, d\ln f_i^{\mathrm{ad}}; \quad (T = \mathrm{const}) \cdot \tag{7.3}$$

Where μ_i^{ad} is the chemical potential in the adsorbate, R is the ideal gas constant, and T is temperature. For defining the amount of the adsorbed mixture, the Gibbs–Duhem equation is applied for constant temperature, and in differential form (Eq. 7.4):

$$\frac{A}{n^{\mathrm{ad}}} = \sum_i z_i \left(\frac{\partial \mu_i^{\mathrm{ad}}}{\partial \Pi}\right)_T \tag{7.4}$$

Where A is the cross-sectional area, and Π is replaced by the area and the so-called spreading pressure. In adsorption, for the adsorbed quantity the concentration is given in loading values for the single components can be derived from Eq. (7.5):

$$q_i = z_i M_i \frac{n^{\mathrm{ad}}}{V^{a^0}} \tag{7.5}$$

Where M_i is molecular mass and V^{a^0} is volume in the bulk fluid.

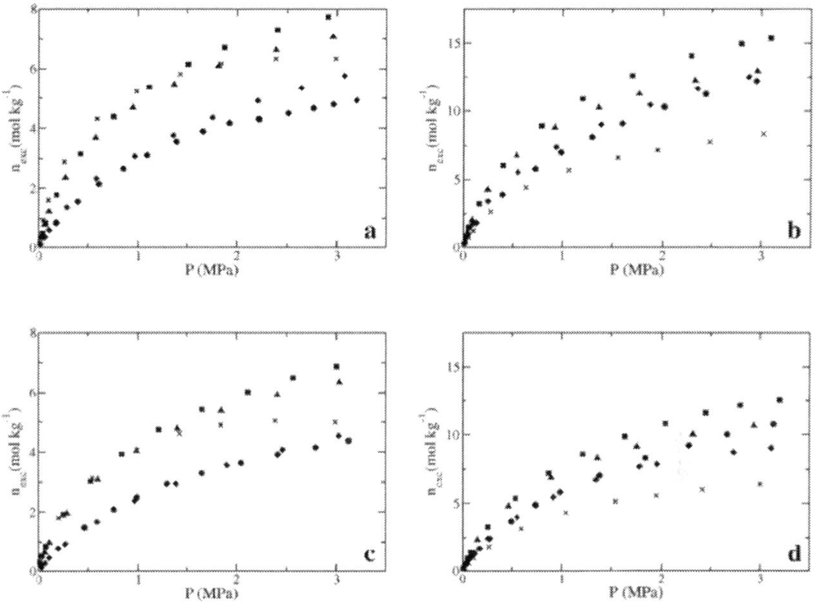

Figure 7.8 Excess adsorption isotherms for samples CNR-115 (circle), RX 1.5 (square), CGran (diamond), Rox 0.8 (triangle), and GAC 1240 (crosses). (A) CH_4 adsorption at 303.15 K, (B) CO_2 adsorption at 303.15 K, (C) CH_4 adsorption at 323.15 K, and (D) CO_2 adsorption at 323.15 K.

An example of experimentally determined adsorption isotherms for CH_4 and CO_2 pure gas were obtained for a series of five microporous activated carbons at two working temperatures 303.15 and 323.15 K (Fig. 7.8). As can be seen in Fig. 7.8A and C displays the methane adsorption isotherms that the sample presents the greater methane values throughout the pressure range is the RX 1.5 activated carbon (square), and Fig. 7.8B and D illustrates the carbon dioxide adsorption isotherms that this sample also presents the greater CO_2 adsorption The results showed that with augmenting temperature, a reduction in the adsorption capacity of both gases was determined for all of the activated carbons; this behavior is characteristic of physical adsorption processes (Peredo-Mancilla et al., 2018).

7.6 SCFs for removal of contaminants

Among current technologies used for removing contamination (Kohli, 2019), SCFE is a comparatively recent and potentially feasible method. Various SCFE processes have been presented for removing intractable organic compounds and toxic from a range of contaminated elements. These comprise the regeneration of adsorbents and soil remediation which is used to treat wastewater streams. As a separation technique for environmental control, SCFE has numerous different benefits over conventional liquid extraction methods and incineration. Interestingly enough, the contaminant is eliminated from the solvent in a concentrated form through a change in pressure or temperature and can be entirely secluded on expansion to atmospheric pressure (Tomasko, Macnaughton, Foster, & Eckert, 1995).

There are different reasons which are included rapid developing, changing technologies, industrial products, practices generated inappropriate disposal, and spill of detrimental materials in the last century threaten public health and the environment. Contamination of soil/sediment by VOCs and POPs including PCBs, PAHs, DDT, dioxins and furans, pesticides, and so forth is still a significant concern. Conventional methods such as thermal desorption, incineration, and liquid solvent extraction are expensive and expose air at risk and result in residual pollution (Anitescu & Tavlarides, 2006).

The SCE method largely has built on information obtained on the use of SCFs for extracting organic compounds from solid substrates for analytical work. Early attempts in the scope of environmental SCE concentrated on studies related to using SCCD to extract a large variety of detrimental chemicals from spiked soil/sediment samples. The next stage was to approve the feasibility of the new technology using contaminated soils and sediments gathered from different sites as real-world/native contaminated samples. Soil/sediment cleanup using SCE processes is available nowadays thank to investigation attempts during the last decades. Dr. Akgerman is a pioneer and carried out extensive work in the domain of environmental SCE. His acute reviews, scientific articles, numerous communications at national and international conferences, seminars, research reports, and so forth. have been adored by the scientific community. He was a pioneer to employ tracer response methods (both pulse and step inputs) to specify equilibrium and kinetic parameters for SCE of contaminants from soils and activated carbon and published for the first time adsorption isotherms for a solute dividing between a solid phase and the SCCD phase. He modeled the dynamics of SCE on soils using orthogonal collocation on finite elements. This work served as an example for subsequent SCE studies involving natural products and polymers (Anitescu & Tavlarides, 2006).

7.6.1 A brief technology description

The treatment technologies can be divided into three major categories: biological, physical/chemical, and thermal. SCE as a subcategory of solvent extraction is classified in the second category. The SCF is selected such that it has a higher affinity for contaminants than for the bulk of the solid substrate. Most of the SCE processes involving SCCD ($T_{cr} = 31°C$; $P_{cr} = 74$ bar) individually or along with cosolvents (entertainers; modifiers) from which the most typical is methanol ($T_{cr} = 239°C$; $P_{cr} = 81$ bar). SCE uses the attractive SCF characteristics (i.e., liquidlike density, low viscosity, high diffusivity, and no surface tension) to extract contaminants at optimal temperature, pressure, and flow rate conditions. Rather than destroying pollutants in situ at relatively low concentrations with considerable energy and/or materials just to be absorbed by the bulk matrix, the SCE process is the first stage of two- or multistage treatment technology. SCE does not annihilate either contaminants or the soil/sediment. The extracted pollutants have high concentrations and can be subsequently destroyed more cost-effectively. All SCE processes for treatment duties use the same unit operations including feed preparation, extraction, separation of solids and solvent, and recovery of the solvent.

- Pretreatment

  A pretreatment is known as a frequently required step for SCE systems. It means soil/sediment preparation for an optimal SCE process. The roots, rocks, and other residues are eliminated and the water content adjusted appropriately. Other features of the soils/sediments may be also adjusted for example pH and chemical composition.

- Treatment

  As has been shown in Fig. 7.9, environmental SCE eliminates the organic matter from contaminated solids followed by solute-solvent separation and solvent recycle. In the SCE stage, the solids that have been excavated are spotted in a pressure vessel and extracted with a stream of SCFs recirculating, which are well mixed with the contaminated matrix to permit contaminants to transfer to the solvent. Then, physical methods have been used to separate the clean matrix, the solvent, and the solute (s). Organic contaminants are exposed to a suitable destruction technology while the clean solvent is recycled to the extraction stage. The soils/sediments are not considerably influenced by an SCE process and can be returned to the natural site.

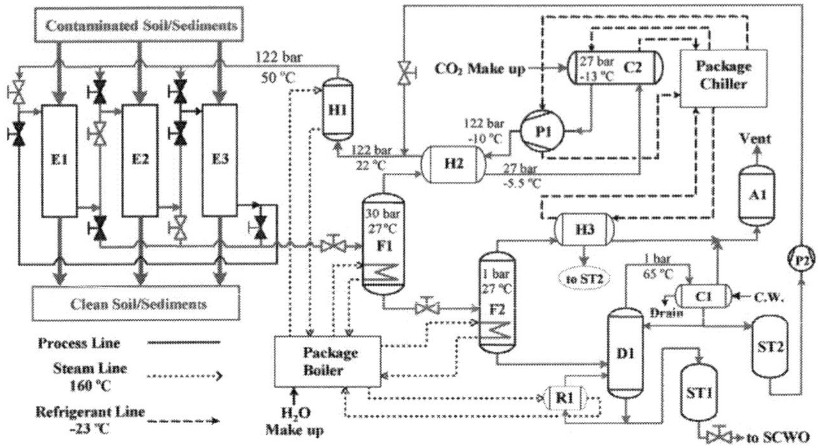

Figure 7.9 A flow diagram for an SCE process to extract PCBs from soil/sediments with SC CO_2–MeOH solvents. The symbols are: *A1*, adsorption column; *C1*, column condenser; *C2*, CO_2 condenser; *D1*, distillation column; *E1*–*E3*, extractors; *F1*–*F2*, flash tanks; *H1*–*H3*, heaters; *P1*–*P2*, pumps; *R1*, column rebuilder; *ST1*, PCB storage tank; *ST2*, MeOH storage tank (Smart et al., 1997).

- Posttreatment

  SCE is not considered as a stand-alone technology. This technology needs to further specific treatment and disposal techniques of the concentrated contaminants and any separated cosolvent and water to complete the cleanup Fig. 7.9. Side streams treatment including spent solvent/cosolvent, spent activated carbon, air emissions, and so forth must also be remediated. However, SCE systems are designed such that work without air emissions.

7.6.2 Solubility studies

SCE projects are planned to prepare engineering design data to develop this process to treat contaminated soils/sediments. The solubility data allows assessing the most favorable SCF from capacity and selectivity views. Solute and SCF structure and thermodynamic properties (i.e., polarity and polarizability, critical point T–P values, vapor pressure, etc.), specific solute–SCF interactions, and so forth affect solute solubility in an SCF. The most important factors influencing the solubility are considered to be solute vapor pressure and SCF density. The solubility amount of a solute in an SCF provides the maximum amount of the solute that can be extracted at an equilibrium state.

Basically, static and dynamic methods are two different approaches that have been used to attain solubility data of various solutes in SCFs. To overcome the main flaws of the conventional methods, a new quasi-static technique has been recently proposed, being categorized between the static and dynamic methods. A very slow SCF flow rate does not considerably perturb thermodynamic equilibrium of the solute—SCF—matrix of subsequent states but allows the attainment of a large number of solubility data-points in a short time (several tens per isotherm).

7.7 SCFs for polymer and plastics industries

Over 100 years, scientists and engineers have been informed that SCF solvents provide the potential for new processing protocols. However, it is only in the past three decades that SCF solvents have been studied or used as solvents in different fields including processing foods, nutraceuticals, and polymeric materials; serve as a reaction medium for polymerization; as environmentally more favorable solvents for solution coatings, powder formation, impregnation, encapsulation, cleaning, crystal growth, and antisolvent precipitation; and as mixing/blending aids for crystalline or viscous materials. By establishing a better understanding of the underlying physics and chemistry of SCF—solute behavior, this wide range of applications could be further extended. Polymerizations and polymer modifications in SCF media have been widely interested in recent years, and this affinity will likely increase due to industrial motivations, such as solvent replacement (Sun, 2002).

All of the examples represented in this chapter are related to the applications of SCF-based technology for processing polymers. The principles section explains a molecular thermodynamic view for interpreting SCF—polymer phase behavior. The following section covers applications consists of three parts. The first part gives observations about the phase behavior of polymers in SCF solvents with some attention to supercritical CO_2. Examples are represented on the effect of polymer structure and solvent quality on solubility. The second part explains the industrial-scale design of Ferro Corporation's SCF-based technology with a focus on engineering details including heat transfer, mixing, and process control schemes. The last part prepares an abstract explanation of several SCF-based applications. Besides special attention toward supercritical CO_2 in this chapter, many of the results can be used to processing with other SCF solvents. Supercritical fluids are currently researched and attracted commercial interest in the polymer industry. They can be used for various processes such as solvents in polymer synthesis, as plasticizing agents, or for chemical recycling. SCFs such as carbon dioxide and water provide a green solution because they do not endanger the earth's environment. From the standpoint of their ever-increasing importance in the polymer industry, there is a requirement to fully understand how SCFs interrelate with polymeric materials in order to fully find out the potential that can be attained from their application (Krukonis, 1988).

The major interest in processing by using $scCO_2$ is the plasticization of polymers, so the effects of plasticization will be discussed. Plasticizers are used as additives in the mixture to change the rheology and/or the mechanical characteristics of the final product. One of the best-known applications of plasticizers is PVC production, where plasticizers can be used to reduce stiffness and lower T_g. Additive affects the polymer depends on its nature and the structure of the polymer. The solubility of the additive will result in dissolving the additive and swelling. The interaction of molten thermoplastic polymers and supercritical fluids is regarded as an alternative for plasticizing agents in the processes. This will important especially when dealing with thermally sensitive polymers with high T_g and also highly viscous materials. In general, plasticization is recognized by changing the polymer system that results in:

- A decline of the glass transition temperature $\left(T_g\right)$
- More plasticity at room temperature
- Enhanced flexibility of individual polymeric chains
- Enhanced toughness

As compared to crystalline regions in semicrystalline polymers, amorphous materials and amorphous regions in semicrystalline materials are more easily penetrated. Also, the crosslinking degree influences the ability of the plasticizer to penetrate. Elastomers that have a less dense network of crosslinks may swell in a solvent but a thermoset with a high crosslink degree neither swells nor dissolves. During the synthesis of polymer, in general, plasticization can be acquired either by adding a small quantity of comonomer causing a disturbance in the subsequent crystallinity and chain packing or by blending a low molecular weight compound with a higher molecular weight polymer. Both techniques result in an increase in the free volume with the subsequent changes in physical properties. Another method to raise plasticization is to use a high-pressure gas, a process called gas-induced plasticization. It is observed that by dissolving $scCO_2$ into a polymer melt the viscosity will be decreased. There is obviously a great deal of interest in trying to process high viscosity melts. It also provides processing temperature-sensitive polymers at lower temperatures, as well as lower processing temperatures for materials with high T_g. As previously mentioned, the effect of $scCO_2$ is swelling of amorphous polymers and they can absorb a greater amount of carbon dioxide than crystalline polymers, and therefore amorphous polymers have a potential for an increase in plasticization and foaming. For materials with high viscosities, a decrease of viscosity can be achieved during processing by an increase in the process temperature. However, an increase in temperature can also result in heat break-down, if it is essential to raise temperatures above the normal operating range. Therefore, it is an extremely attractive alternative that could reduce the viscosity of polymers. This is not only about the drop in viscosity and therefore improvements in the inability to process, and also the ability to decrease the processing temperatures. This enables a reduction in energy consumption and therefore results in more cost-effective processes. Obviously, this is a very attractive proposition for material processors.

The high pressures needed for the containment of SCF within the polymers are similar to those already encountered in processing operations such as injection molding and extrusion. Therefore, it is not a big problem to make system modifications for use with an SCF such as $scCO_2$. Also, reducing the pressure can be easily used to eliminate the gas. In the following, processes developed for use with supercritical fluids will now be discussed with the main focus on the foam applications. Energy savings as the other major benefit of using supercritical fluids should also be borne in mind throughout (Krukonis, 1988).

7.7.1 Polymers in supercritical carbon dioxide ($scCO_2$)

There are some reasons for making carbon dioxide (CO_2) the most commonly used SCF including its low critical temperature, low toxicity, and high purity at a low cost. Moreover, CO_2 is the second most abundant and the second least expensive solvent on the planet. Being in the gas state under ambient conditions facilitates its removal from polymeric products, thus it will be cost-effective in other secondary operations such as drying and solvent removal (Lindy, 2015).

As compared with many SCF materials, CO_2 has a relatively low operating pressure and temperature, which makes achieving supercritical conditions less expensive. For example, it presents in a supercritical state is often working the range of manufacturing processes such as extrusion or injection molding have been already experienced by polymers. It can also be readily removed by simply decreasing the pressure. Easy recovery for recycling is another advantage and it can be used to replace hazardous or toxic materials such as Freon. It can be dissolved with a variety of organic solvents, and also be used to replace some organic solvents. With a few notable exceptions, most of the high molecular weight polymers are not soluble in this material, an example is fluoropolymers which do dissolve. However, polymers can take up a considerable amount of the material in a supercritical state. Amorphous polymers have different behavior from semicrystalline materials, this difference will be discussed in the following section (Lindy, 2015).

At near the critical conditions, amorphous materials have a greater absorbance of CO_2 than semicrystalline polymers and a greater amount of plasticization occurs. This is because of the absorbance of CO_2 only into amorphous regions and not crystalline regions. As the fluid concentrates, a phase transition from glass to liquid may occur due to sorption and swelling in amorphous polymers (or amorphous regions of semicrystalline materials). The glass transition temperature $\left(T_g\right)$ of the polymer falls down significantly, a property that can be exploited in polymer processing. Amounts in the range of 10%–30% w/w can be absorbed by these materials and treatment with just 8%–10% can decrease a T_g of 80°C to below room temperature. An increase in temperature and pressure results in an increase of absorption amount, however, two factors of polarity and the structure of the polymer also specify the solubility of any particular polymer being used. Plasticization of the polymer also results in an increase in crystallinity. It increases the melting point and induces changes in the mechanical characteristics of materials. The different response to $scCO_2$ infers that amorphous and semicrystalline polymers may be more appropriate for some applications than others. Definitely, within the domain of this review, it is amorphous polymers that are of most interest. Numerous attempts have been made to quantify the plasticization effects of $scCO_2$ on different polymers. However, there are some obstacles to definitely state exact levels of either fluid uptake or level of plasticization effect for any given polymer, conditions, and types of the same material which have been tested. Because of these factors, any comparison is virtually impossible. Despite this, however, in the domain of polymer processing, the effectiveness of supercritical technology has been already established. Supercritical fluids diffuse into polymers more easily than larger liquid solvents due to their small size.

This property makes the controlled rapid sorption of SCF easier through the pressure and temperature adjustment. The SCF has been used for polymer plasticization impacts polymer processing activities including separation and fractionation of polymers for extraction of impurities, membrane conditioning, impregnation and extraction of impurities, foams, gels, and fibers, and polymerization reactions and the production of microparticles. Changes in the physical and mechanical properties of polymers are readily observed by the sorption of supercritical CO_2, with the consequent effect of the decline of T_g. The major effects of supercritical carbon dioxide on amorphous polymers are shown in Fig. 7.10 and Table 7.9 represents the summarized effects of $scCO_2$ on both amorphous and semicrystalline polymers.

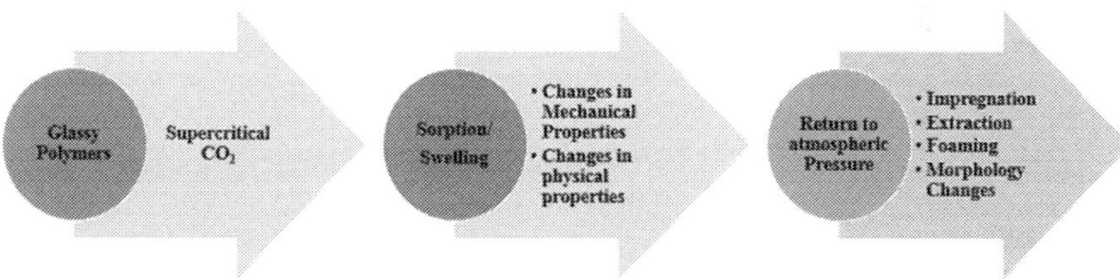

Figure 7.10 Effect of supercritical carbon dioxide on amorphous polymers. *From: Lindy, J. (2015). Supercritical fluid extraction: Technology, applications and limitations. Nova Science).*

Table 7.9 Summary of the effects and applications of $scCO_2$ on amorphous and semicrystalline polymers.

Semicrystalline	Amorphous	
Gas absorption	Low (only in amorphous regions)	High
Plasticization effect	Low (only in amorphous regions)	High
Crystallinity	Small increases may be seen (only in amorphous regions)	Increased
Common examples	Nylon 66, PVDF, PP, HDPE	ABS, HIPS, PC, PMMA, PET
Potential applications	Replacements for CFCS in cleaning applications (limited applications as amorphous dependent on the degree of crystallinity)	Foaming and production of microcellular structures, impregnation, Surface modification, introduction of crystallization

7.8 SCFs for recovery of heavy metals from wastewater

In the past two decades, due to the potential applications of SCF extraction technology, it has been interested considerably by chemists and engineers as an environmentally friendly solvent for chemical processing (Cansell et al., 1998). Attention toward supercritical fluids, particularly carbon dioxide as a solvent for use in extraction processes, has been motivated by increased environmental authorities limiting the usage of conventional solvents. This has caused the development of both analytical and large scale techniques for organic waste extraction from contaminated. Extensive literature is available describing the solubility and extraction of organic compounds using supercritical CO_2 in detail (Smart et al., 1997).

Metal waste contamination highly threatens the environment and human health. Several methods are used to monitor and clean-up metal contaminated soils. Time-consuming traditional extraction methods, such as BCR, Tessier, etc., have been used for decades that are not environmentally acceptable. Moreover, analytical difficulties are confronted when employing successive extraction methods. In addition, significant quantities of environmentally unacceptable chemicals are involved in these methods (Yabalak & Gizir, 2013).

SCE is a good alternative to conventional solvent extraction, not only to eliminate heavy metals, but also the morphology and structure matrix is maintained. Carbon dioxide and water are recognized as environmentally acceptable solvents because of their properties including low cost, and nontoxic and nonflammable nature and solubility characteristics, which are adjustable by varying the temperature and pressure. During recent years, the wide majority of SCFE's have concentrated on the usage of supercritical carbon dioxide ($scCO_2$) to extract metals as some metal complex, due to the low toxicity of carbon dioxide and it is environmentally acceptable.

7.9 Industrial upscaling

Numerous important research and development means are dedicated to applications of SCFs in industry, some mercantile plants are now operating or understructure and a small number of firms have obtained some know-how in process upscaling in SCF adsorption and extraction design. Several upscaling standards for SCF can be detected in essays. Nevertheless, the studies are often conflicting and inconclusive. Hence, more studies in this field are required. In the case of industrial upscaling, Vardanega et al. in a most comprehensive review of the main SCF technologies used for food processing, including transformation and extraction, presented positive results (Vardanega, Náthia-Neves, Veggi, & Meireles, 2019).

Most SCF factories have separators designed as large empty gravity colonists, but remarkable savings could be made by using many smaller-volume cyclonic chambers that are of special interest. In the case where the substrate is very fugacious and notably soluble in the fluid at the separation stage or present at trace concentration, sorption onto an adsorbent bed is needed to either refine the fluid before recycle, for instance removing the pesticides from raw material or to recover a valuable particle of the extract that should have been lost in the other respect, like light ends of flavors or fragrances. In such cases, granular activated carbon is often prior by virtue of inexpensive and has a very large sorption capacity (~ 1 g/g) and collection efficiency simultaneously it can be easily desorbed by the fluid itself at SCF states. In most cases, it is also precious to filter the fluid before recycling in order to avoid imprisonment of raw material or extract fine particles that may amass and plug inside the condenser, and finally necessitating to stop the factory for a difficult-to-operate cleaning (Clavier & Perrut, 2004).

7.10 Economic evaluation/justification

The development of more efficient and economical chemical transformation processes is still a significant challenge. There is a requirement to make the organic synthesis reactions in the polymer, chemical, and pharmaceutical industries highly selective, economical, and environmentally friendly. To developed such processes, the application of SCFs as reaction solvents has been increasingly investigated in the last 2 decades due to the potential preponderances offered by these media over conventional liquid solvents and gaseous diluents (Mendes & Coelho, 2005).

The total processing cost is the sum of the total operating cost and the depreciation of the plant for over 10 years. Capital cost factors include expenses for installation, engineering and construction, and contingency funds and fees. Appropriate cost factors are utilized for materials, utilities, and maintenance. The total treatment cost including depreciation is $175/m^3 of soils/sediments for the base case by using the costs mentioned above. Economic analysis of the SCE discloses that the operating costs of the proposed process are in the range of $122–$154/m^3 of soils/sediments. By comparison of these costs with the costs of other alternative technologies, it is

Table 7.10 Comparison of relevant processes for remediation of contaminated soils/sediments.

Process	Advantages	Disadvantages	Cost ($/m^3)
Supercritical fluid extraction	High extraction efficiency; no second hand pollution; soil structure is left intact	Further treatment of the extracts; high-pressure equipment; soil excavation required	122–154
In situ thermal desorption	No excavation required; applicable to heterogeneous soil; broad applicability	Further treatment of the extracts; high temperatures, high energy; a slow process	100–380
Liquid solvent extraction	High extraction efficiency; low T–P conditions; portability	Difficult solute separation; further treatment of the extracts; residual pollution	250–1169
Incineration	High destruction efficiency; continuous flow system; broad applicability	High cost and energy consumption; emission of toxic gas/ash; irreversible soil degradation	>> 360
Bio-remediation	No second hand pollution; broad land applications; ambient conditions	Low destruction efficiency; very slow process; inactive for some solutes	245–474

suggested the SCE process is an economically promising and attractive treatment for PCB contaminated soils/sediments (Table 7.10, Anitescu & Tavlarides, 2006).

7.11 Future opportunities and challenges

SCFs are solvents for the future by virtue of an ecological view, supercritical fluid-based technologies propose important benefits over organic solvent technologies. Consequently, high environmental concerns and increases in organic solvent prices require new solvents have augmented. Also, the requirements for novel products with unique characteristics, lower energy, and high purity utilization by means of low process temperatures for applications in all fields of human activity are increasing. By using SCFs, such products can be produced (Cocero, 2018; Knez et al., 2019).

7.12 Conclusion

Process intensification is defined as increasing the efficiency of operating units. One of these ways is to use supercritical fluids. SCFs have lured much interest as sorption solvents for well over several decades. In this chapter, at first, we were discussed SCF and properties. SCFs have some different profits as extraction solvents for environmental management such as the ability to remove contaminants. The SCFs have attracted much attention in terms of their adjustable density, viscosity, low toxicity, and low cost. Adsorption equilibrium must be considered in order for SCF technology to become economically competitive. Therefore, we can conclude that advances in the field of SCFs and high-pressure technologies have opened up new potential for components and products gained with inexpensive and environmentally friendly methods.

References

Agrawal, R., Belemkar, S., & Bonde, C. (2020). Supercritical fluid chromatography versus liquid chromatography for the enantiomeric separation of itraconazole. *Microchemical Journal, 158*, 105320.

Anitescu, G., & Tavlarides, L. L. (2006). Supercritical extraction of contaminants from soils and sediments. *The Journal of Supercritical Fluids, 38*, 167–180.

Berk, Z. (2018). *Food process engineering and technology*. Academic press.

Brunner, G., & Johannsen, M. (2006). New aspects on adsorption from supercritical fluid phases. *The Journal of Supercritical Fluids, 38*, 181–200.

Cansell, F., Rey, S., & Beslin, P. (1998). Thermodynamic aspects of supercritical fluids processing: Applications of polymers and wastes treatment. *Revue de l'Institut Français du Petrole, 53*, 71–98.

Clavier, J.-Y., & Perrut, M. (2004). In *Scale-Up issues for supercritical fluid processing in compliance with GMP*.

Cocero, M. J. (2018). Supercritical water processes: Future prospects. *The Journal of Supercritical Fluids, 134*, 124–132.

Fédoroff, M., & Lefèvre, G. (2006). Sorption mechanisms and models. Their influence on transport calculation. In B. J. Merkel, & A. Hasche-Berger (Eds.), *Uranium in the environment: Mining impact and consequences*. Berlin, Heidelberg: Springer Berlin Heidelberg.

Goodship, V., & Erich, O. O. (2005). *Polymer processing with supercritical fluids*. iSmithers Rapra Publishing.

Herrero, M., Cifuentes, A., & Ibañez, E. (2006). Sub-and supercritical fluid extraction of functional ingredients from different natural sources: Plants, food-by-products, algae and microalgae: A review. *Food Chemistry, 98*, 136–148.

Kazarian, S. G. (2000). Polymer processing with supercritical fluids. In , *42. Polymer science series CC/C of vysokomolekuliarnye soedineniia* (pp. 78–101).

Knez, Ž., Pantić, M., Cor, D., Novak, Z., & Hrncic, M. K. (2019). Are supercritical fluids solvents for the future? *Chemical Engineering and Processing - Process Intensification, 141*, 107532.

Kohli, R. (2019). Chapter 6 - applications of supercritical carbon dioxide for removal of surface contaminants. In R. Kohli, & K. L. Mittal (Eds.), *Developments in surface contamination and cleaning: Applications of cleaning techniques*. Elsevier.

Krukonis, V. J. (1988). Processing with supercritical fluids: Overview and applications. *Supercritical Fluid Extraction and Chromatography: Techniques and Applications, 366*, 26–43.

Lindy, J. (2015). *Supercritical fluid extraction: Technology, applications and limitations*. Nova Science).

Mendes, M. F., & Coelho, G. L. V. (2005). Desorption processes: Supercritical fluid regeneration of modified clays. *Adsorption, 11*, 139–146.

Monteil-Rivera, F., Fedoroff, M., Jeanjean, J., Minel, L., Barthes, M. G., & Dumonceau, J. (2000). Sorption of selenite (SeO(3)(2-)) on hydroxyapatite: An exchange process. *Journal of Colloid and Interface Science, 221*, 291–300.

Myers, A. L., & Prausnitz, J. M. (1965). Thermodynamics of mixed-gas adsorption. *AIChE Journal, 11*, 121–127.

Peredo-Mancilla, D., Hort, C., Jeguirim, M., Ghimbeu, C. M., Limousy, L., & Bessieres, D. (2018). Experimental determination of the CH_4 and CO_2 pure gas adsorption isotherms on different activated carbons. *Journal of Chemical & Engineering Data, 63*, 3027–3034.

Sanchez-Camargo, A. de P., Mendiola, J. A., Elena, I., & Herrero, M. (2014). *Supercritical fluid extraction*.

Smart, N. G., Carleson, T., Kast, T., Clifford, A. A., Burford, M. D., & Wai, C. M. (1997). Solubility of chelating agents and metal-containing compounds in supercritical fluid carbon dioxide. *Talanta, 44*, 137–150.

Sun, Y.-P. (2002). *Supercritical fluid technology in materials science and engineering: Syntheses: Properties, and applications*. CRC Press.

Sunol, A. K., & Sunol, S. G. (2001). Substitution of solvents by safer products and processes. *Nitric Oxide, 93*, 64.0.

Tomasko, D. L., Macnaughton, S. J., Foster, N. R., & Eckert, C. A. (1995). Removal of pollutants from solid matrices using supercritical fluids. *Separation Science and Technology, 30*, 1901–1915.

Vallejos-Burgos, F., Ohba, T., & Kaneko, K. (2019). Fundamental aspects of supercritical gas adsorption. In K. Kaneko, & F. Rodríguez-Reinoso (Eds.), *Nanoporous materials for gas storage*. Singapore: Springer Singapore.

Vardanega, R., Náthia-Neves, G., Veggi, P. C., & Meireles, M. A. A. (2019). 3 - supercritical fluid processing and extraction of food. In F. Chemat, & E. Vorobiev (Eds.), *Green food processing techniques*. Academic Press.

Weibel, G. L., & Ober, C. K. (2003). An overview of supercritical CO_2 applications in microelectronics processing. *Microelectronic Engineering, 65*, 145–152.

Yabalak, E., & Gizir, M. A. (2013). Subcritical and supercritical fluid extraction of heavy metals from sand and sewage sludge. *Journal of the Serbian Chemical Society, 78*, 1013.

Yonker, C. R., & Smith, R. D. (1990). Sorption isotherms of mobile phase components in capillary supercritical fluid chromatography. *Journal of Chromatography A, 505*, 139–146.

Index

Note: 'Page numbers followed by "*f*" "*t*" refer to figures and tables respectively.'

N

O

P

R

S

Kolibri 360 GmbH
Ferdinand-Jühlke-Straße 7
99095 Erfurt, Deutschland
produktsicherheit@kolibri360.de